Kalahari Hyenas

Kalahari Hyenas

Comparative Behavioral Ecology of Two Species

M. G. L. Mills

Kalahari Hyenas: Comparative Behavioral Ecology of Two Species

ISBN-10: 1-930665-83-0
ISBN-13: 978-1-930665-83-5

Library of Congress Control Number: 2003107803

THE BLACKBURN PRESS
P. O. Box 287
Caldwell, New Jersey 07006
U.S.A.
973-228-7077
www.BlackburnPress.com

Dedicated to
the memory of Stevie

Contents

Foreword

This beast has a stone in its eye, also called an Yena, which is believed to make a person able to foresee the future if he keeps it under his tongue. It is true that if an Yena walks round an animal three times, the animal cannot move. For this reason they affirm that it has some sort of magic skill about it. (Translated from Latin, from the twelfth century Bestiary in the University of Cambridge Library.)

Perhaps it is not merely fortuitous that of all the African animals, the ancients should have selected the hyaena as a vehicle for magical powers. Of course, there are solid scientific explanations for this; the animals' nocturnal habits around human habitation, the consumption of people's mortal remains, the spectacular hermaphroditic appearance, the uncanny similarity between the calls of a spotted hyaena and the utterances of deranged humanity. But apart from all rational explanations for the strange hold of the animal over people, there is a magic about hyaenas which can only be understood by those who have watched them for some time. There is a now growing band of us, who came to the African bush with all our prejudices, with all that 'common knowledge' about hyaenas which proved so totally wrong, and who just fell for the spell of animals which were so totally different.

The hyaena family is the smallest one of the carnivores, with a membership of only three species of proper hyaenas, and the somewhat aberrant aardwolf. But many more hyaenas existed in days before our time; we are only seeing the tail end of hyaena glory. Some were huge, and, for instance, skulls of over 90 cm long have been found in North Africa. If only we had been there to see the packs of those huge, probably rhinoceros-sized animals roaming the plains, driving alongside in our vehicle to spot the details of their hunting behaviour, and comparing them with all the smaller species!

On a smaller scale, that is just what Gus Mills did with the present-day remainder of the hyaenas. The two species which now live alongside each other in the Kalahari provide a beautiful opportunity for a comparison between the solitary and the gregarious, between a specialist and an opportunist. At the same time, it takes some superb fieldcraft to realize that opportunity in the harsh, almost desert conditions. And fieldcraft is something which Gus has built into him, something which is not likely to show itself in his scientific writing, which can only be appreciated when watching the man in his natural habitat, the Kalahari.

Gus knows the southern Kalahari like a hyaena does, at night and by day; the endless dune ridges, hundreds of miles of sand, scrub and grass; to

an outsider it is all the same and intensely hostile. This book will not tell of
the thrills and alarms of the midnight cross-country race, following packs of
hyaenas, taking dune tops at speed in the pitch dark, in the hope that the
other side will be a gentle slope. All this with the one aim of keeping up
with the study, with the hyaena ahead, with its way of staying alive. In this
way Gus, and his wife Margie, put an immense investment, in time and
energy, into the acquisition of background knowledge and skills to get the
observations needed for a book like this.

The study is outstanding, especially in the amount of detailed knowledge
amassed over such a long period, of individual hyaenas and their offspring,
and of the offspring of offspring. The interactions within clans, the
differences between the position of individual brown and spotted hyaenas
within their respective social environments, the social and feeding
strategies of each, all this within the overwhelming Kalahari habitat with its
complicated ecology: Gus has proved it to be a goldmine for the behavioural ecologist.

The results of the painstaking nightwork reported here will not be the
last study of hyaenas; certainly, more is to come. But it will be a unique
building stone in our understanding of carnivore behaviour in general, and
of hyaena social organization and ecology in particular. It will help in
designing management strategies for a species which needs protection
rather than persecution, the brown hyaena. And in its comparisons with
spotted hyaenas elsewhere, it will show the fascinating flexibility of the
other species, ostensibly a specialist.

Hyaenas do not have to circle their prey three times in order to catch it;
Gus Mills showed that perhaps the ancients were wrong on that. But
through his stunning pictures, his solid data and interpretations, he also
showed that the general idea of the hyaenas behind those medieval
statements was pretty well right: animals of bewitching attraction, with a
profound contribution to make to behavioural ecology.

Hans Kruuk
Banchory, Scotland

Acknowledgements

I received so much help and encouragement during the fieldwork and in the writing of this book and it is a pleasure to acknowledge the contributions of all concerned. If I have forgotten anybody, I apologize, and assure them that I do appreciate their help, even if my memory lets me down.

To my employers, the National Parks Board of South Africa and the past and present directors Dr Rocco Knobel, Mr Dolf Brynard, and Dr Tol Pienaar, I extend my sincere appreciation for giving me the privilege to live and work in the Kalahari Gemsbok National Park for 12 years, for financially supporting nearly every aspect of the study, and for giving me permission to publish this book. I also gratefully acknowledge the Department of Wildlife and National Parks, Botswana for allowing me to work in the Gemsbok National Park.

Besides the National Parks Board, financial support for the brown hyaena study was provided by the S.A. Nature Foundation, the Wildlife Society of southern Africa, the Coca-Cola Export Corporation, the Council for Scientific and Industrial Research, and the Endangered Wildlife Trust. The late Mr Jack Rivett-Carnac provided four drop-door traps and the late Dr Reay Smithers lent me a starlight scope and provided valuable advice on the trapping of brown hyaenas and other aspects of the fieldwork. Russel Freidman and Eddie Bourhill kindly transported various drafts of the manuscript from Johannesburg to London.

Professors Fritz Eloff and Koos Bothma, University of Pretoria, were instrumental in my going to the Kalahari and in securing funds before I became employed by the National Parks Board. They showed sustained interest in the project for its entire duration.

I was extremely fortunate to meet Dr Hans Kruuk early on in my study. He provided invaluable guidance in the field, generously shared his ideas, and was a constant source of encouragement during various periods of writing. The few short visits he made to the Kalahari were highlights of inspiration. He read the entire manuscript, making many constructive comments. I am also most grateful to him for writing the foreword.

Many colleagues in the National Parks Board assisted and cooperated during the study in many ways. Elias le Riche, Warden of the Kalahari Gemsbok National Park, provided much logistic support. The bushmen trackers Ooi and Houtop laboured for many hours on the spoor of brown hyaenas, and later my assistant Hermanus Jaeggers did sterling work tracking spotted hyaena spoor and accompanying me on many field trips and nocturnal expeditions. His enthusiasm and love for hyaenas was infectious. These three men taught me much about the bush and bushcraft. In the preparation of the manuscript I was greatly assisted by Peter Retief with statistics and computer work, Lorna Stanton and her assistant Lazarus Makitla with the photographs, Heather Wildi who expertly drew the figures, assisted in part by Jessica Hughes, and Merle Whyte and Maureen Rochat who did much of the typing. I am also grateful to Dr Gerrie de Graaff, Dawie de Villiers, Rian Labuschagne, Dr Anthony Hall-Martin, the late Stoffel le Riche, Doempie le Riche, Isak Meyer, Dr Piet van der Walt, and Piet van Wyk for help in many ways. Michael Knight kindly allowed me to use some of the observations he made on spotted hyaenas after I left the Kalahari.

Dr David Macdonald, the series editor, spent many hours discussing hyaenas and other carnivores with me and reading thoroughly through two earlier drafts of the manuscript, making many valuable and constructive comments. I am most grateful to him for his time and effort, as I am to other colleagues who have helped with the preparation of the final manuscript. Dr Martyn Gorman helped with Chapters 4, 5 and 7, provided the three-dimensional maps, and contributed directly to the sections on scent marking and home range and movement patterns. Dr Philip Richardson read Chapters 1 to 4, and we have also discussed many aspects of hyaena behavioural ecology. Jack da Silva and Paul Stewart commented on most of the manuscript, and Shaughn Doolan read Chapter 4. I have also benefitted from discussions with Professors John Skinner, Jan Nel and Rudi van Aarde, and Drs Valerius de Vos, Joe Henschel, Lawrence Frank, Francios Messier, Gustav Peters, Rolf Peterson, Butch Smuts and Roy Bengis, and Richard Goss.

Permission to reproduce my previously published material in original or modified form was kindly granted by *Koedoe* (Figs 2.1, 2.4, 2.5, 2.6, 2.18, 2.19, 2.21, 4.4, and Tables 2.18, 2.21, 2.22), *Journal of Zoology* (Figs 4.14, 4.18, 4.22a, 4.23a, 4.24, 5.20, 5.21, 5.22, 5.23), *Nature* (Figs 3.19, 3.23, 3.28, 4.6), *South African Journal of Wildlife Research* (Fig. D.1 and Table 8.3), *South African Journal of Zoology* (Figs 5.8, 5.9, 5.12 and Tables 5.12, 5.13), *Animal Behaviour* (Tables 5.2, 5.4, 7.4, 7.5, 7.6), *Zeitschrift für Tierpsychologie* (Figs 1.1, 1.4, 2.8, 3.9, 3.10, 3.12, 3.13, 3.38a, 7.1, 7.4 and Tables 3.5, 3.11, 7.2), *Behavioural Ecology and Sociobiology* (Fig. 4.2 and Tables 4.4, 7.1, 7.3), and American Society of Mammalogists (Fig. 5.5).

Last, but not least, I extend my sincere thanks to my family for their loyal support. My parents and parents-in-law provided logistic support and encouragement. My wife and co-worker, Margie, not only assisted in much of the fieldwork, she also contributed significantly to the analysis of the data, typed a large part of the manuscript, and above all provided the love and moral support needed to complete this work.

M. G. L. Mills
Skukuza

1 Introduction

1.1 The study

This book is about the behavioural mechanisms which are relevant in the ecology of two species of hyaena, the brown hyaena* and the spotted hyaena, in a near pristine environment, the semi-arid southern Kalahari.

The study of spotted hyaenas in East Africa by Kruuk (1972) was a classic in the field of behavioural ecology, and drastically changed people's conception of the spotted hyaena. Kruuk showed that, far from being a solitary scavenger, having little impact on the ecosystem, it can be as formidable a predator as any of Africa's large carnivores. He also discovered that spotted hyaenas have a complex and flexible social system. But perhaps the greatest contribution of this study was to show that an animal's behaviour is inexorably tied to its diet and the distribution of its food supply.

Kruuk (1975a, 1976) and Kruuk & Sands (1972) then studied other members of the Hyaenidae in East Africa, the striped hyaena and the aardwolf. These animals were found to have very different life-styles from the spotted hyaena and from each other. This too was mainly brought about by differences in diet.

Since then the influence of diet and food dispersion on behaviour and social organization has been studied in other carnivores (see Macdonald 1983, Gittleman & Harvey 1982, Bekoff *et al.* 1984 and references therein), and, specifically, in the brown hyaena by Owens & Owens (1978) and Mills (1982a). These studies have further shown the great amount of intra- and interspecific flexibility in carnivore social behaviour, brought about by the pervasive influence of food.

Major developments in behavioural ecology since Kruuk's hyaena studies have included an appreciation of the importance of kinship within social groups, and of the variation in individual behaviour, particularly with regard to reproductive strategies. This was largely due to the influence of Wilson's (1975) book *Sociobiology*. As a result, there has recently been an emphasis on studying the behaviour of individuals, rather than classes of individuals or groups. Amongst carnivores the studies of Bertram (1976), Bygott *et al.* (1979), Packer & Pusey (1982, 1983a, b), Davies & Boersma (1984), Hanby & Bygott (1987), Packer *et al.* (1988), and Pusey & Packer (1987) on lions have been particularly illuminating. The importance of kin relationships amongst brown hyaenas was shown by Mills (1982b) and

---

*Scientific names of all plant and animal species mentioned in the text can be found in Appendix A.

Owens & Owens (1984), but was not appreciated in spotted hyaenas (Bertram 1979) until more recently (Mills 1985a, Frank 1986a & b). Useful reviews of developments in these fields in other carnivores can be found in Macdonald & Moehlman (1982), Macdonald (1983) and Bekoff et al. (1984).

These are interesting and theoretically important aspects of behavioural ecology. However, it was the conservation of the hyaenas within a framework for the conservation of the entire southern Kalahari ecosystem that was the main stimulus for the study. The brown hyaena is listed in the (IUCN) *Red Data Book* of endangered species, where its status is given as vulnerable. It has a limited distribution, being essentially an inhabitant of the southwest, arid and adjacent drier parts, of the southern savanna biotic zones of Africa (Von Richter 1972, Smithers 1983), with the southern Kalahari a stronghold for this species. The spotted hyaena's range has become drastically reduced in southern Africa this century (Kruuk 1972, Smithers 1983). Within the southern Kalahari it was believed that their numbers had dropped in the 20 years preceding the study (C. le Riche, personal communication). The aims of the study, therefore, were: firstly, to assess the roles of the two species of hyaena in the southern Kalahari ecosystem, and the factors and mechanisms responsible for limiting their populations; and, secondly, to contribute to our understanding of carnivore behavioural ecology.

Within this framework, the objectives of the study can be divided into two broad categories; those pertaining to the feeding ecology and social systems of the two species separately, and those pertaining to comparisons between the species. In the first category the important questions to ask pertaining to feeding ecology are: What are the diets of the two species? How do they procure their food? How is their food distributed? What effect do they have on their prey populations, and what are their relationships with the other carnivores, particularly with regard to competition? Pertaining to their social systems it is important to know how the social systems are related to the food supply and other environmental factors, how they are maintained through behaviour, and what the roles of individuals in hyaena societies are – specifically, how individuals enhance their reproductive success.

The important comparative questions which are posed are: how are these two closely related species adapted to inhabit the same area? How has the influence of diet affected their foraging behaviour, social systems, communication patterns, denning behaviour, and mating systems, and what has been the influence of a common phylogenetic origin on these facets? What is the relationship between the brown hyaena and spotted hyaena, and what effect do they have on each other's populations?

Much of the data on the brown hyaena presented here have been previously published (Mills 1978a, 1982a, b, c, 1983a, b, Mills & Mills 1978, 1982, Mills et al. 1980, Gorman & Mills 1984), as have some on the spotted hyaena (Mills 1985a, Mills & Gorman 1987). In addition, comparisons between the two species have been made in previous publications (Mills 1978b, 1984a, 1989, Mills & Mills 1977). In this volume I have synthesized this previous work, with the addition of much unpublished data, particularly on the spotted hyaena. Articles from which data are drawn are mentioned at the beginning of each chapter, but are only referred to in the main text if a fact or conclusion is stated without presenting the data.

1.2   The study area

The Kalahari is a large, sand-filled basin in the west of the south African subcontinent, covering nearly one third of the area and forming what is probably the largest sand-veld area in the world. Kalahari sand stretches from 28°S to 1°N and covers an area of some 1 630 000 km^2 (King 1963). In the south lie the Kalahari Gemsbok (South Africa) and adjacent Gemsbok (Botswana) National Parks, which together make up an area of 36 190 km^2 (Fig. 1.1). It is this area that is referred to as the southern Kalahari. The western, southwestern, and southern boundaries of the national parks are fenced off from stock-farming areas; the remaining boundaries are open and border on controlled hunting areas (Von Richter & Butynski 1973).

   In the southern Kalahari the sand deposits are arranged in a series of long, parallel dunes, which are fixed by vegetation. Four rivers drain the area, two from the northwest and two from the east. These rivers are usually dry, only flowing for short periods in abnormally wet years. The two from the northwest, the Auob and the Nossob, run through the national parks, the Nossob forming the boundary between the Kalahari Gemsbok and Gemsbok National Parks (Fig. 1.1). Irregularly scattered in the dunes are numerous shallow, flat depressions or pans, relict lakes of a wetter epoch some 16 000 years ago (Lancaster 1979).

   The southern Kalahari is a semi-desert region having an irregular rainfall and experiencing large temperature fluctuations, both on a daily and seasonal basis. In summer (October to April) maximum daily temperatures are normally 30–40°C, dropping to 10–20°C at night. In winter (May to September) the maximum daily temperature is usually around 20°C, dropping to 5°C or lower (as low as −10°C) at night (Anon. 1986). Most of the area falls within the 200–300 mm rainfall isohyets, and 70% or more of the rain usually falls between January and April. At Nossob camp, in the centre of the study area (Fig. 1.1), the mean annual rainfall between 1972 and 1983 was 255.4 ± 44.2 mm (range 106.5–602.5 mm), the years 1972–1977 being generally wet years and the years 1978–1983 being dry. Evaporation rates are high, so that rainfall is unlikely to exceed evaporation for any month of the year. Consequently, free-standing water is rare, only being found temporarily on pans and along river beds after excep-

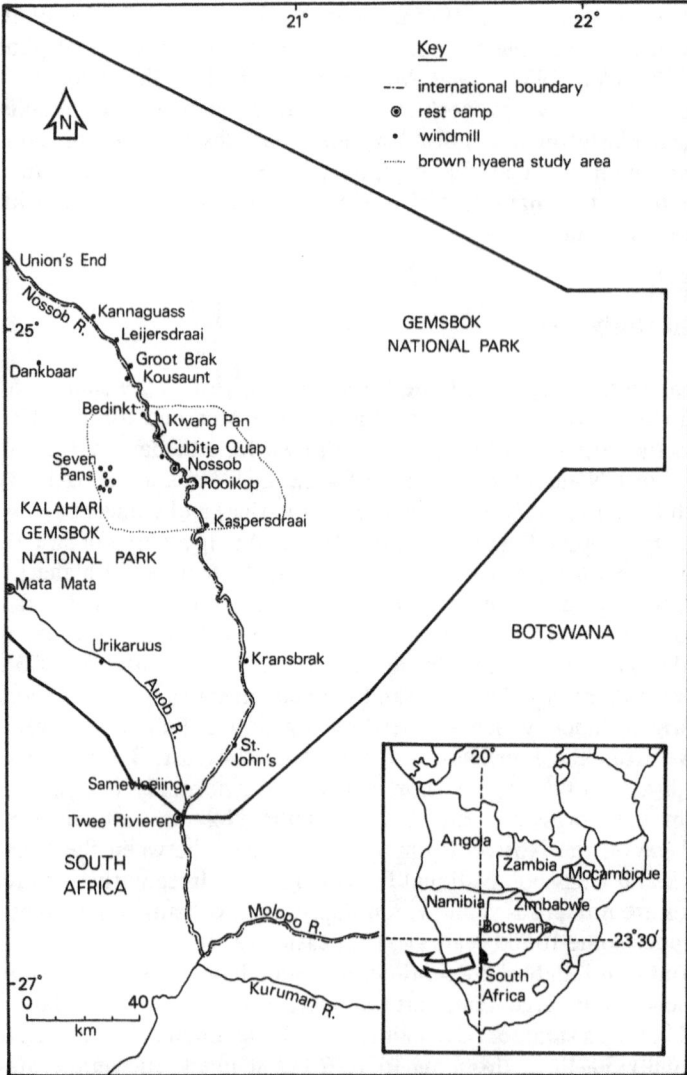

Figure 1.1 The southern Kalahari, showing the Kalahari Gemsbok and Gemsbok National Parks and other places mentioned in the text, as well as the brown hyaena study area.

tionally heavy rains. However, water from boreholes is available to game at roughly 5–10 km intervals along the river beds, and less regularly in the dunes of the Kalahari Gemsbok National Park.

The vegetation is described by Acocks (1975) as the western form of the Kalahari thornveld, an extremely open shrub or tree savanna. More detailed descriptions of the vegetation can be found in Leistner (1959a & b,

1967), Blair Rains & Yalala (1972), Leistner & Werger (1973) and Van Rooyen *et al.* (1984). For the purposes of this study two main habitats are recognized; the river beds and immediate environs (river habitat), and the adjacent dune areas including pans (dune habitat) (Fig. 1.2).

The former, which make up less than 5% of the area, are dominated by large camelthorn and grey camelthorn trees, the tall shrub, blackthorn, perennial grasses such as buffalo grass and speckled vlei grass and, after good rains, by the annual grasses feather-top chloris and Kalahari sour grass (Fig. 1.3). Large limestone plains up to 4 km wide, dominated by the dwarf shrub rhigozum (colloquially known as 'driedoring') and the perennial small bushman grass, form a significant part of this habitat type along the Nossob river.

The dune areas are dominated by smaller camelthorns and grey camelthorn trees, as well as by shepherd's tree, and by tall perennial grasses such as Lehmann's love grass, gha grass, giant stick grass, and dune grass (Fig. 1.4). Pan surfaces are bare; the vegetation on the fringes resembles that of the limestone flats along river beds, with rhigozum and small bushman grass as well as another dwarf shrub, monechma, dominating.

In addition to insectivores, bats and small rodents, 36 mammal species have been recorded from the area (Smithers 1971, personal observations). The ecological distribution of the ungulates and relevant smaller animals is discussed in Chapter 2. Besides the two hyaena species, other large

Figure 1.2 Parallel sand dunes and the Nossob river bed with road running along it; an example of the two main habitat types in the southern Kalahari.

carnivores regularly found in the area are lions, leopards and cheetahs, with African wild dogs being extremely rare.

Figure 1.3 A herd of gemsbok in the Nossob river bed near Bedinkt.

Figure 1.4 Typical dunes habitat near Rooikop. Tall perennial grasses with widely scattered trees and shrubs.

1.3 Methods

The openness of the southern Kalahari vegetation makes it possible to observe large animals such as hyaenas, which quickly become habituated to a vehicle. I was, therefore, able to follow study animals for long periods in a four-wheel-drive vehicle, even at night, relying on the moonlight, the parking lights of the vehicle, or a hand-held spotlight for illumination. Care was taken not to use a spotlight to dazzle the prey when hyaenas were hunting.

Most observations of brown hyaenas were on individuals with radio-collars. The radio-collars shortened the time it took to find an animal, and made it possible to quickly relocate it if visual contact was lost. Two beta-lights (sealed glass capsules coated with a green phosphor and filled with tritium gas, causing them to glow) were fitted to each collar (Fig. 2.4) which helped in maintaining visual contact with moving brown hyaenas. After fitting two radio-collars with beta-lights onto spotted hyaenas, it became clear that it was possible to locate them on a regular basis and to follow them adequately, without the use of this expensive equipment.

A second useful feature of the southern Kalahari is the layer of sand covering most of the region. This makes it possible to track spoor and to reconstruct aspects of the animals' behaviour (Eloff 1964). However, interpreting the spoor can be difficult and the method has limitations. It was more useful for spotted hyaenas than for brown hyaenas, as the latter forage solitarily, move erratically, and often consume a food item entirely.

The two species were intensively studied during two slightly overlapping periods. The brown hyaena study period was between 1972 and 1980, and most observations were carried out in an area of 2750 km^2 centred around Nossob camp (Fig. 1.1). I made intensive long-term observations on a clan called the Kwang clan, backed up by less intensive observations on five other clans. In all I spent over 2500 h observing brown hyaenas and, with the help of trackers, followed about 1200 km of their spoor.

Apart from opportunistic observations on spotted hyaenas during the brown hyaena study, regular observations on this species were made from January 1979, through until February 1984. Since then certain aspects (see Chapter 7) have been monitored by M. Knight, and I made a short visit to the area in December 1984. The spotted hyaena study area covered the entire Kalahari Gemsbok National Park, plus a strip along the Nossob river bed extending for 30 km into the Gemsbok National Park. I singled out one group of spotted hyaenas, the Kousaunt clan, for intensive long-term observations, made less intensive observations (mainly from tracking spoor) on three others, the Seven Pans, Kaspersdraai, and St John's clans, and kept notes on several others. I spent in excess of 1500 h observing spotted hyaenas and tracked spoor for nearly 3500 km.

The mapping of movements of both species was achieved by noting the vehicle's odometer reading at each change in direction. Direction was occasionally determined by means of a hand-held compass (held 10 m away from the vehicle), but normally I recorded one of 16 compass divisions by using the sun, moon, or stars as reference points. The odometer reading at known points such as windmills or roads was also recorded. Each movement was plotted on a 1:10 000 map (see Appendix D).

Throughout the study it was necessary to immobilize hyaenas to fit radio-collars, or to clip their ears for identification. Most brown hyaenas were caught in a baited drop-door trap and then immobilized. As suggested by Smuts *et al.* (1977a & b), dragging the bait one or two kilometres either side of the trap increased the capture success rate. In 666 trap-nights 29 different brown hyaenas were caught 169 times, as were six lions and four leopards once each. The main drawback of this technique was that it was non-selective; some individuals were 'trap-happy' and others were 'trap-shy'. For example, four brown hyaenas of the Kwang clan were caught 17, 11, 8, and 0 times respectively in 160 trap-nights between January and December 1978. Trap-shy brown hyaenas, and all spotted hyaenas, were darted with a Cap-Chur pistol or rifle from a stationary vehicle.

Three types of immobilizing drug were used; phencyclidine hydrochloride, ketamine hydrochloride (these two in combination with the tranquillizer combellum), and a 1:1 combination of the immobilizing drug tiletamine hydrochloride and the tranquillizer zolazepam hydrochloride (CI-744. Phencyclidine hydrochloride has the drawback that the animals take a long time to recover, even with dosage rates as low as 0.5–1.0 mg/kg, and is not recommended for use on hyaenas. Dosage rates of 10–20 mg/kg ketamine and 5–10 mg/kg CI–744 gave satisfactory results.

All individuals caught were placed into one of five age-classes based on tooth eruption and wear, particularly with regard to the wear on the third premolar in the lower jaw, after Kruuk (1972) and Mills (1982c). Animals in age-class 1, i.e. until they received their full permanent dentition at about 15 months, were regarded as cubs. Hyaenas between 15 and 30 months were regarded as sub-adults, after which they were classed as adults. Observations on some known-aged animals revealed that hyaenas pass from age-class 2 into age-class 3 in their fourth year and into age-class 4 at about six years of age. Kruuk (1972) recorded that East African spotted hyaenas pass into age-class 5 at 16 years of age.

At certain places in the text, particularly in Chapter 7, certain individual hyaenas are identified by anthropomorphic names. I have done this to help the reader remember important individuals and thus to follow the text. Wherever possible, behaviour described in the text is illustrated with black and white photographs. These were mainly taken at night and therefore suffer from high contrast. In addition many have been made from colour transparencies. In spite of the loss of quality experienced because of these difficulties, it was felt worthwhile including them to give the reader a visual image of the behaviour.

Unless otherwise stated, statistical means are given with standard error where possible. As in most behavioural studies, non-parametric statistical tests were mainly used (Siegal 1956), with the few parametric tests taken from Sokal & Rholf (1969). Statistical tests are given in figures and tables where possible, otherwise they are listed at the end of each chapter with a figure in brackets given in the relevant place in the text, after Clutton-Brock *et al.* 1982. Statements in the text that two samples differ indicate a significant difference at the 0.05 level or higher.

1.4   The animals

1.4.1   Evolution and phylogenetic relations

The Hyaenidae are a small family today, comprising three genera and four species; there are two members of the genus *Hyaena* (striped hyaena and brown hyaena), one *Crocuta* (spotted hyaena), and one *Proteles* (aardwolf). The hyaenids are unique in their development of bone-crushing teeth, except for *Proteles* which has a reduced dentition.

The earliest known fossil of a hyaenid is a small viverrid-like creature known as *Progenetta* from early European Miocene deposits (Beaumont 1967 in Savage 1978). However, early hyaenid fossils are few and the exact evolutionary lines and migratory routes taken by the family have not been established. The Hyaenidae reached their peak in terms of number of species in Africa during the early Pleistocene, from which nine species have been recorded (Ewer 1967). Ewer (1973) points out that this period corresponds with an abundance of sabre-tooths and artiodactyls. Sabre-tooths were highly specialized meat-eaters, and it is likely that the presence of a primary predator of this type opened a niche for bone-crushing specialists. The disappearance of the sabre-tooths during the middle Pleistocene corresponds with the shrinking of the hyaena fauna.

The *Crocuta* and *Hyaena* lines have been separated since the late Miocene (Thenius 1966, Howell & Petter 1980). The origin of the spotted hyaena has not been established, although it probably originated in Africa in the early Pleistocene, having come into Africa as an ancestral form (Ewer 1967). During the second half of the Pleistocene the genus *Crocuta* was widespread, with only a single species in Africa, but several in Eurasia, some much larger than the extant form and others considerably smaller (Ewer 1967). Today the spotted hyaena is found over Africa south of the Sahara, except for the Congo Basin and most of South Africa, where it has recently disappeared (Kruuk 1972).

The two extant *Hyaena* species probably have a common origin in southern Africa in the late Miocene (Howell & Petter 1980), making them one of the longest lived carnivore genera (Savage 1978). A subspecies of the striped hyaena *H. hyaena makapani* is known from this area, probably from the upper Miocene (Howell & Petter 1980). The roots and dispersals of *Hyaena* are unknown. Today the brown hyaena is found in southern Africa and the striped hyaena extends from north and east Africa, through Arabia and Asia Minor to India (Rieger 1981).

The origins of *Proteles* are also unknown. Hendey (1974) favours the hypothesis that it stemmed from a group of late Miocene hyaenids called ictitheres, which seem to be fairly similar to *Hyaena* (Howell & Petter 1980). Thenius (1966) even suggested that *Proteles* might be more closely related to *Hyaena* than *Hyaena* is to *Crocuta*. Genetic studies have also shown the close relationship between *Proteles* and the extant hyaenas (Wurster & Benirschke 1968, Seal 1969, Von Ullrich & Schmitt 1969, Wurster 1969). The aardwolf occurs in two discrete populations in Africa, one in southern Africa, the other from central Tanzania through to northeastern Sudan (Smithers 1983).

1.4.2   Physical characteristics

The brown hyaena is smaller than the spotted hyaena (Table 1.1), and is dark brown to black in colour with a white neck-ruff, long hair along the back and pointed ears (Fig. 2.5). The forequarters are well developed in comparison with the hindquarters, resulting in a characteristic spoor where the front foot is considerably larger than the hind foot. The upper molar is two or more times the size of the first upper premolar (Smithers 1983). The southern Kalahari data show little size difference between males and females, athough the total length of males is significantly greater than that of females (Table 1.1) (1).* In agricultural areas in the Transvaal, Skinner & Ilani (1979) found males to be heavier than females. Moreover, these males at 47.1 kg, are heavier than Kalahari ones (2), although females at 42 kg are not (3).

The spotted hyaena is a yellowish colour with dark brown to black spots on the body, short hair and round ears (Fig. 2.11). Southern Kalahari females in particular become brown as they grow older, and lose most of their spots. The hindquarters are also less developed than the forequarters, but not to the extent that they are in the brown hyaena. Consequently, the difference in size between the forefoot and hindfoot is not so marked in the spotted hyaena as it is in the brown hyaena. The upper molar is greatly reduced, either much smaller than the first upper premolar, or absent.

Female spotted hyaenas are generally thought to be larger than males (see Hamilton *et al.* 1986 and references therein), although this has been disputed by Whateley (1980) and also to an extent by Hamilton *et al.* (1986). My data show that in terms of mass in particular, but also heart girth, southern Kalahari females are significantly larger than males, although they are not longer (Table 1.1) (4). This was also found to be the

* Figures in brackets refer to statistical tests at the end of the chapter.

Table 1.1   Means and standard deviations of some body measurements of brown hyaenas over 30 months of age and spotted hyaenas over 36 months of age.

	Brown hyaena		Spotted hyaena	
	Male	Female	Male	Female
Mass (kg)	40.2 ± 3.0	37.6 ± 3.4	59.0 ± 2.8	70.9 ± 3.3
Total length (mm)	1466 ± 78.7	1399 ± 60.1	1585 ± 63.9	1631 ± 73.5
Heart girth (mm)	816 ± 36.2	782 ± 51.2	885 ± 64.0	995 ± 42.8

case in the Masai Mara National Reserve, Kenya (Hamilton *et al.* 1986), and in the Kruger National Park, South Africa (Van Jaarsveld *et al.* 1988). Heart girth is more strongly correlated with mass than is total length in brown hyaenas (Mills 1982c), and size is a product of length and girth. Although female spotted hyaenas may not be longer than males, they have a larger girth and are heavier. In addition, females do not reach full size until they are over three years of age (Matthews 1939, personal observations), whereas males do so at about 30 months. Asympototic weight is only reached at approximately ten years (Van Jaarsveld *et al.* 1988). Studies showing no significant difference in mass between male and female spotted hyaenas may have included adult females that were not full-grown.

1.4.3   Density estimates

It is difficult to census nocturnal animals such as hyaenas, particularly the elusive brown hyaena, therefore I have resorted to two different, indirect methods to measure the densities of the two species in the southern Kalahari.

The brown hyaena density was estimated as follows: the average territory and group (adults and sub-adults) sizes ± 95% confidence limits for brown hyaenas in the study area were calculated from data presented in Chapter 4. By extrapolation, three estimates of the number of group-living brown hyaenas in the Kalahari Gemsbok and Gemsbok National Parks combined were made; one using average group size and territory size, one using the upper 95% confidence limit for group size and the lower one for territory size, to give a theoretical maximum, and vice versa for the third to give a theoretical minimum. Observations on marked animals suggested that 64% of the population comprised group-living individuals, the remainder of the population being itinerants (section 4.3.1). Therefore 36% of each of the three estimations of the number of group-living brown hyaenas was added to itself, in order to obtain a figure for the total number of brown hyaenas in the area. From these figures the average density was 1.8 brown hyaenas/100 km^2 (651 animals) with the 95% confidence limits being 0.4–4.4 brown hyaenas/100 km^2.

Spotted hyaenas are not as evenly distributed over the southern Kalahari as brown hyaenas. They are more common in the Kalahari Gemsbok National Park and bordering areas of the Gemsbok National Park than they are further to the east, probably because of the influence of the river beds on their prey (Chapters 2, 4) and the provision of water (section 8.2.1). My density estimates for this species, therefore, are for the Kalahari Gemsbok National Park only.

Spotted hyaenas are more conspicuous than brown hyaenas, and I am confident that I knew of all the clans in the Kalahari Gemsbok National Park. Apart from the four clans that were studied in detail, there were six others in the area in the early 1980s. The composition of six of the ten clans was reasonably well known (section 4.2.1); four contained about ten adults and sub-adults each, and two contained about five each. The sizes of the other four clans were unknown and are assumed to have been between five and ten individuals each. This gives a figure of 70–90 clan-living hyaenas in the Park. As with the brown hyaena, a proportion of the spotted hyaena population are itinerants (section 4.3.2). The percentage they form is unknown, but they are less common than in the brown hyaena population. I have arbitrarily assumed that it was 10%. This increases the figure to between 77 and 99 spotted hyaenas in the Kalahari Gemsbok National Park, at a density of $0.8–1.0/100\,\mathrm{km}^2$, less than half that of the brown hyaena density.

1.5 Summary

1. The aims of the study are: (a) to describe the niches of the brown hyaena and spotted hyaena in the southern Kalahari ecosystem; (b) to study the factors and mechanisms which limit their populations; (c) to assess the roles of individuals in hyaena societies; and (d) to contribute to the management of the two species.
2. The southern Kalahari is a pristine, arid, sand-veld region of dunes fixed by vegetation, with two dry river beds cutting through the area.
3. Hyaenas were observed mainly at night from a four-wheel-drive vehicle, or, in the case of spotted hyaenas particularly, by tracking spoor.
4. The Hyaenidae comprises four extant species from three genera. The *Crocuta* and *Hyaena* lines have been separated since the late Miocene. *Hyaena* are one of the longest-lived carnivore genera.
5. The brown hyaena is smaller than the spotted hyaena. Male and female brown hyaenas do not differ much in size, whereas spotted hyaena females are larger than males.
6. The density of brown hyaenas in the southern Kalahari was calculated at $1.8/100\,\mathrm{km}^2$ and the spotted hyaena density in the Kalahari Gemsbok National Park was $0.8–1.0/100\,\mathrm{km}^2$.

Statistical tests

1. Comparison of mean body measurements of male ($N = 11$ for weight, $N = 9$ for others) versus female ($N = 9$) brown hyaenas from the southern Kalahari.
    (i)   *Weight*. Student's t test: $t = 1.765$; d.f. $= 18$; $p = 0.08$
    (ii) *Total length*. Student's t test: $t = 2.030$; d.f. $= 16$; $p = 0.06$
    (iii) *Heart girth*. Student's t test: $t = 1.627$; d.f. $= 16$; $p = 0.12$.
2. Comparison of the mean weights of male brown hyaenas from Transvaal agricultural areas ($N = 8$) versus the mean weights of males from the southern Kalahari ($N = 11$).
    Student's t test: $t = -6.12$; d.f. $= 17$; $p < 0.001$; two-tailed.
3. Comparison of the mean weights of female brown hyaenas from Transvaal agricultural areas ($N = 4$) versus the mean weights of females from the southern Kalahari ($N = 9$).
    Student's t test: $t = -2.07$; d.f. $= 11$; $p = 0.06$; two-tailed.
4. Comparison of mean body measurements of male ($N = 9$) versus female ($N = 7$) spotted hyaenas from the southern Kalahari.
    (i)   *Weight*. Student's t test: $t = 7.64$; d.f. $= 14$; $p < 0.001$
    (ii) *Total length*. Student's t test: $t = 1.40$; d.f. $= 14$; $p > 0.05$
    (iii) *Heart girth*. Student's t test: $t = 2.17$; d.f. $= 14$; $p = 0.05$.

2 Feeding ecology

Food and its acquisition are vital to the life histories of all anima
Individuals must acquire sufficient nutrients, not only for survival, but a
to reproduce. Indeed, the main theme in this book is an investigation of
influence of diet and food dispersion on the behaviour of hyaenas.

It is, therefore, necessary to describe the dispersion pattern of food a
the diets of the hyaenas in detail before describing their behaviour. I
only must hyaenas compete with each other for limited food resources, t
also compete with other carnivorous animals in the southern Kalah
which utilize the same resources. The diets of lions, leopards and cheeta
the amount of dietary overlap and relationships between the two hya
species and the other carnivorous animals, therefore, are all relevant.

2.1 Food availability

Discussion on the feeding ecology of an animal necessitates knowledge
the availability of its food. For predators the distribution and abundance
the prey (here mainly the antelope, but for the brown hyaena some of
smaller animals as well), as well as aspects of their population dynam
such as breeding success and sex ratios are relevant. For scavengers i
important to know about mortality factors of potential food species.

2.1.1 Large and medium-sized antelope

2.1.1.1 ECOLOGICAL DISTRIBUTION OF ANTELOPE

Apart from the two smallest and sedentary antelope species, the comm
duiker and the steenbok, the antelope community of the southern Kalah
comprises five species of nomadic or semi-nomadic, medium-sized to la
species which are important to hyaenas – gemsbok, blue wildebeest,
hartebeest, eland and springbok. The kudu is so rare in the study area a
be of little importance.

The unpredictable rainfall in the southern Kalahari has led to a ma
nomadic existence for the antelope. Most rain usually falls in late sum
(January–April), when springbok, wildebeest, hartebeest and gems
concentrate along the dry river beds (Fig. 2.1). As conditions dry out,
antelope move out of the river beds, gemsbok and hartebeest more rap
than springbok and wildebeest. Remnant populations of these latter
species are always to be found along the river beds (Fig. 2.1);
springbok because of their ability to switch their diet from grazing
selective browsing, the wildebeest mainly because of the provision

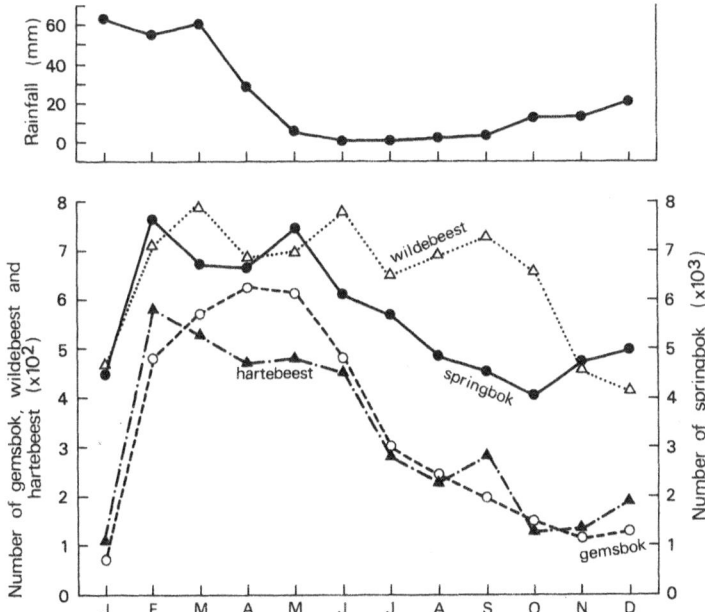

Figure 2.1 Average monthly rainfall and average monthly abundance of ungulates along the river beds, 1972–1982.

Table 2.1   Annual mean number of ungulates per count along a 40km strip of the Nossob river bed north of Nossob camp.

Year	Springbok	Gemsbok	Hartebeest	Wildebeest	Eland	Steenbok
1974	583	17	46	40	0	0
1975	411	30	35	28	0	0
1976	1098	59	74	124	0	0
1977	909	64	111	172	0	1
1978	492	22	66	194	0	2
1979	609	17	48	176	0	1
1980	532	24	44	128	0	1
1981	329	6	17	66	0	1
1982	812	4	10	93	0	0
1983	549	1	6	88	0	0
Mean ± SD	632±237	24±22	46±32	111±58	-	1±1

potable water along the river beds (Mills & Retief 1984a and b). For springbok, gemsbok and hartebeest, but not wildebeest, there was a significant positive correlation between rainfall and average annual abundance along the river beds (Bothma & Mills 1977, Mills & Retief 1984a). Eland rarely come into the river beds.

Table 2.2   Annual mean monthly numbers of ungulates counted per 40km in the dune areas in the vicinity of Nossob camp.

Year	Springbok	Gemsbok	Hartebeest	Wildebeest	Eland	Steenbok
1974	0	3	5	0	14	2
1975	0	12	2	0	0	4
1976	0	7	5	0	0	6
1977	4	6	5	0	0	5
1978	0	10	0	0	0	4
1979	1	27	11	26	0	11
1980	1	32	1	0	1	5
1981	no data					
1982	4	16	3	0	3	7
1983	6	24	2	6	8	8
Mean ± SD	2±3	15±10	4±3	4±9	3±5	6±3

In the dunes the vast area and lack of any obvious pattern in the movements of the antelope, make it difficult to document their numbers and distribution. Generally, the dunes in the spotted hyaena study area (see section 1.3), supported a far lower biomass of ungulates than did river habitat (see examples in Tables 2.1 and 2.2). However, at certain times large concentrations of certain species occurred in the dunes: in September–October 1979 an estimated 172 000 wildebeest swarmed into the spotted hyaena study area, compared with the mean figure of 1896 in this area in other years (see Appendix B); over 13 000 eland and 18 000 hartebeest were estimated to be in the area in September 1981, whereas six months later there was a count of only 12 eland and 259 hartebeest. Estimates of the number of some of the ungulates in the spotted hyaena study area are given in Appendix B.

2.1.1.2 NOTES ON SOME ASPECTS OF ANTELOPE POPULATION DYNAMICS
Gemsbok. Gemsbok have an extended calving season (Fig. 2.2). The adult and sub-adult sex ratio was close to parity and the average percentage of cows with calves under one year of age was 35% (Table 2.3). Predation, almost exclusively by lions (section 2.4.1), accounted for approximately two-thirds of adult mortality (Fig. 2.3).
Wildebeest. Wildebeest have a short calving season, nearly all calves being dropped in December and January. The adult sex ratio was biased towards females and the overall cow to calf ratio was slightly higher than for gemsbok (Table 2.3). Adult mortality was equally divided between predation and non-violent categories (Fig. 2.3). In the dry years 1978–1982, non-violent mortality was particularly prevalent amongst the wildebeest (Mills 1984b).
Springbok. Springbok births usually peak in January, although in some

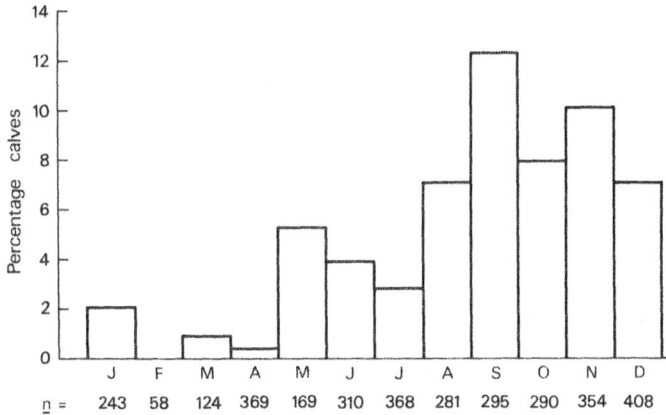

	J	F	M	A	M	J	J	A	S	O	N	D
n =	243	58	124	369	169	310	368	281	295	290	354	408

Figure 2.2 Percentage of red (4–6 months) gemsbok calves counted each month, 1979–1983.

Table 2.3 Some population characteristics of gemsbok, wildebeest, springbok and hartebeest in the southern Kalahari from counts carried out between 1978 and 1983. Calves are under one year old, lambs are under six months old. Sub-adults are 1-2 years old for all species except springbok which are 6-12 months old. It was impossible to differentiate between sub-adult and adult gemsbok.

	Gemsbok $n = 538$	Wildebeest $n = 12\ 342$	Springbok $n = 33\ 164$	Hartebeest $n = 893$
Per cent adult males	38.5	23.6	24.5	22.6
Per cent adult females	45.7	46.8	41.6	45.9
Per cent sub-adults	-	11.0	12.3	11.1
Per cent calves/lambs	15.8	18.6	21.7	20.4
Adult males:adult females	1:1.19	1:1.98	1:1.70	1:2.03
Binomial (two-tailed)	$p > 0.05$	$p < 0.001$	$p < 0.001$	$p < 0.001$
Adult females:calves/lambs	1:0.35	1:0.40	1:0.52	1:0.44

years they may be earlier, and in 1981 lambing occurred from January to April with no peak. Lambs may also be dropped at other times of the year, particularly in April–May and September–October. During the birth season the breeding females congregate on lambing grounds, and in the first month of life lambs are particularly prone to predation (section 2.4.5). Adult females outnumbered adult males in the population, and the average ewe to lamb ratio was higher than for the other three species (Table 2.3). At the peak lambing seasons the average ewe to lamb ratio was 1:0.64. Predation was the most important recorded cause of mortality for springbok adults (Fig. 2.3).

Hartebeest. Most hartebeest calves were born between September and November, with a smaller peak in April and May. There was a preponder-

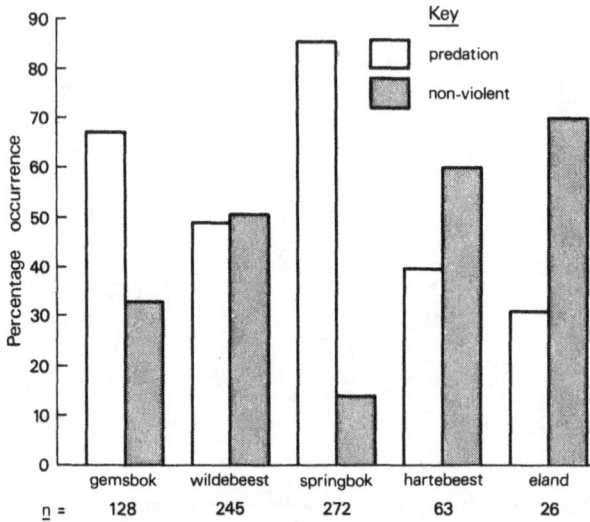

Figure 2.3 Relative incidence of predation and non-violent mortality for various ungulates, 1972–1982.

Table 2.4   Frequency of occurrence of the tracks of different animals in ten strips of 200 x 1.5m located on the sides of the Nossob river bed and in the dunes. Each strip was swept clean of spoor and checked the next day.

Animal	Side of river bed $n = 20$	Dunes $n = 80$	χ^2 d.f. = 1
Steenbok	13(65.0%)	71(88.8%)	8.60; $p < 0.01$
Hare	7(35.0%)	53(66.3%)	7.88; $p < 0.01$
Yellow mongoose	5(25.0%)	52(65.0%)	12.14; $p < 0.001$
Korhaan species	1 (5.0%)	51(51.8%)	24.54; $p < 0.001$
Springhare	8(40.0%)	34(42.5%)	0.21; $p > 0.05$
Cape fox	7(35.0%)	31(38.8%)	0.32; $p > 0.05$
Black-backed jackal	11(55.0%)	22(27.5%)	4.30; $p < 0.05$
Duiker	0 (0.0%)	22(27.5%)	8.74; $p < 0.01$
Bat-eared fox	3(15.0%)	16(20.0%)	0.69; $p > 0.05$
African wild cat	5(24.0%)	9(11.3%)	1.50; $p > 0.05$
Striped polecat	1 (5.0%)	8(10.0%)	1.29; $p > 0.05$
Porcupine	1 (5.0%)	3 (3.8%)	0.15; $p > 0.05$
Ground squirrel	2(10.0%)	2 (2.5%)	0.80; $p > 0.05$
Caracal	0 (0.0%)	3 (3.8%)	2.60; $p > 0.05$
Suricate	0 (0.0%)	2 (2.5%)	2.58; $p > 0.05$
Aardvark	0 (0.0%)	2 (2.5%)	2.58; $p > 0.05$
Pangolin	0 (0.0%)	1 (1.3%)	3.09; $p > 0.05$
Small-spotted genet	0 (0.0%)	1 (1.3%)	3.09; $p > 0.05$
Honey badger	0 (0.0%)	1 (1.3%)	3.09; $p > 0.05$
Tortoise	0 (0.0%)	1 (1.3%)	3.09; $p > 0.05$

ance of adult females in the hartebeest population, and, overall, a slightly higher cow to calf ratio than in the other two large ungulates (Table 2.3). Non-violent mortality accounted for 60% of the adult mortality (Fig. 2.3). *Eland.* Little information on age composition and sex ratios of eland was obtained. Small calves were seen between September and December. Non-violent mortality accounted for 69% of all recorded mortality (Fig. 2.3).

2.1.2   Small animals

The frequency of occurrence of some small animals was measured by track counts (Table 2.4), which revealed the importance of the dunes for many of them. Of the more important species in the brown hyaena's diet (see Tables 2.5 and 2.7), steenbok, hare, and korhaans (two species of ground-nesting birds) showed a clear preference for dune areas (see also Tables 2.1 and 2.2 for steenbok), whereas black-backed jackals preferred the river habitat, and bat-eared foxes and springhares showed no habitat preference. Springhares, however, prefer the hard ground of the river bed itself to the softer ground on the sides, where the track counts in river habitat were done. Night counts with a spotlight revealed an average of 69.4 ± 23.4 springhares per 20 km along the river bed and 25.6 ± 5.9 per 20 km in the dunes.

2.2   Brown hyaena diet

2.2.1   Overall diet

Table 2.5 records all food items brown hyaenas ate while being followed, showing the varied nature of their diet. It is comparable with the observations on spotted hyaenas in Table 2.10. Mammal remains are the brown hyaena's most important food (Fig. 2.4), although the frequency with which they ate different kinds of mammal varied depending on availability. For example, the frequency with which members of the Kwang clan fed on the remains of large versus medium-sized mammals varied (Table 2.6) with the relative abundance of this food along the Nossob river bed (Table 2.1). Wildebeest, in particular, became an important food source for these brown hyaenas during 1978–1980, when non-violent mortality (i.e. starvation, thirst, or disease), due to drought, took a toll on the wildebeest population (section 2.4.1).

Of the 205 (25.8%) unidentifiable food items in Table 2.5, 80% were small items eaten in less than 1 min, and were probably not bone fragments, as I could usually hear these being cracked. Faecal analyses revealed that 60.1% of adult faeces contained insect remains (mainly Coleoptera and Isoptera) and 19.8% contained reptile remains (Mills &

Figure 2.4 A brown hyaena scavenges a hartebeest carcass. Note the radio-collar with mounted beta-light.

Figure 2.5 A large brown hyaena cub opens up a tsama.

Figure 2.6 A brown hyaena with a scavenged secretary bird.

Figure 2.7 A large brown hyaena cub feeds on the remains of its mother which had died a week earlier. Note the adult's radio-collar.

Table 2.5   Brown hyaena diet from direct observations. Reptiles and insects are underrepresented - see text.

Species	No. of carcasses from kills and scavenging	No. of hyaenas eating from kills and scavenging	No. of carcasses from kills only	No. of hyaenas eating from kills only
Mammalia	**329 (41.4%)**	**561 (54.5%)**	**6 (21.4%)**	**12 (35.3%)**
Wildebeest adult	19 (2.4%)	95 (9.2%)	0 (0.0%)	0 (0.0%)
Springbok adult	55 (6.9%)	77 (7.5%)	0 (0.0%)	0 (0.0%)
Springbok lamb	6 (0.8%)	10 (1.0%)	1 (3.6%)	4 (11.8%)
Gemsbok adult	20 (2.5%)	58 (5.6%)	0 (0.0%)	0 (0.0%)
Gemsbok calf	2 (0.3%)	2 (0.2%)	0 (0.0%)	0 (0.0%)
Hartebeest adult	17 (2.1%)	53 (5.2%)	0 (0.0%)	0 (0.0%)
Hartebeest calf	1 (0.1%)	1 (0.1%)	0 (0.0%)	0 (0.0%)
Steenbok	13 (1.6%)	29 (2.8%)	0 (0.0%)	0 (0.0%)
Eland adult	5 (0.6%)	18 (1.7%)	0 (0.0%)	0 (0.0%)
Bat-eared fox	4 (0.5%)	15 (1.5%)	1 (3.6%)	4 (11.8%)
Springhare	5 (0.6%)	12 (1.2%)	2 (7.1%)	2 (5.9%)
Black-backed jackal	3 (0.4%)	7 (0.7%)	0 (0.0%)	0 (0.0%)
African wild cat	3 (0.4%)	7 (0.7%)	0 (0.0%)	0 (0.0%)
Small rodent	4 (0.5%)	4 (0.4%)	1 (3.6%)	1 (2.9%)
Brown hyaena	1 (0.1%)	2 (0.2%)	0 (0.0%)	0 (0.0%)
Lion	1 (0.1%)	1 (0.1%)	0 (0.0%)	0 (0.0%)

	Col 1	Col 2	Col 3	Col 4
Duiker	1 (0.1%)	1 (0.1%)	0 (0.0%)	0 (0.0%)
Cape fox	1 (0.1%)	1 (0.1%)	0 (0.0%)	0 (0.0%)
Suricate	1 (0.1%)	1 (0.1%)	0 (0.0%)	0 (0.0%)
Pangolin	1 (0.1%)	1 (0.1%)	0 (0.0%)	0 (0.0%)
Striped polecat	1 (0.1%)	1 (0.1%)	1 (3.6%)	1 (2.9%)
Skin and bones	165 (20.8%)	165 (16.0%)	0 (0.0%)	0 (0.0%)
Aves	**29 (3.7%)**	**32 (3.1%)**	**6 (21.4%)**	**6 (17.6%)**
Ostrich adult	9 (1.1%)	9 (0.9%)	0 (0.0%)	0 (0.0%)
Ostrich eggs*	7 (0.9%)	8 (0.8%)	-	-
Korhaan	5 (0.9%)	5 (0.5%)	4 (14.3%)	4 (11.8%)
Korhaan eggs	1 (0.1%)	1 (0.1%)	-	-
Others	7 (0.9%)	9 (0.9%)	2 (7.1%)	2 (5.9%)
Reptiles	**7 (0.9%)**	**7 (0.7%)**	**0 (0.0%)**	**0 (0.0%)**
Insects**	**16 (2.0%)**	**16 (1.6%)**	**16 (57.1%)**	**16 (47.1%)**
Fruits**	**183 (23.0%)**	**183 (17.8%)**	-	-
Gemsbok cucumber	78 (9.8%)	78 (7.6%)	-	-
Tsama	73 (9.2%)	73 (7.1%)	-	-
Others	32 (4.0%)	32 (3.1%)	-	-
Miscellaneous**	**25 (3.1%)**	**25 (2.4%)**	-	-
Unidentified	**205 (25.8%)**	**205 (19.9%)**	-	-
Total	**794**	**1029**	**28**	**34**

* One nest and six single eggs
** Number of patches fed on
*** Faeces, human wastage, etc.

Table 2.6   Average distances moved between different food items eaten by brown hyaenas of the
Kwang clan during three time periods, and χ^2 values (d.f. = 1) for the differences, taking
into account the distance they were followed during each period.

Item	1975 vs. 1976	1976 vs. 1977-78	1975 vs. 1977-78
Large mammal remains	36km vs. 49km χ^2 =0.91; p>0.05	49km vs. 19km χ^2 =16.21; p<0.001	36km vs. 19km χ^2 =4.59; p<0.05
Medium-sized mammal remains	100km vs. 24km χ^2 = 11.03; p<0.001	24km vs. 66km χ^2 =11.58; p<0.001	100km vs. 66km χ^2=0.77; p>0.05

Mills 1978). As so few of these food items were seen to be eaten by brown
hyaenas, it is likely that insects and reptiles formed the bulk of the
unidentified food items listed in Table 2.5.

Wild fruits (Fig. 2.5) are also important foods for brown hyaenas, and
other vertebrates (Fig. 2.6), birds' eggs (Fig. 3.10), as well as virtually any
edible food, are eaten. For example, an old female killed at Nossob camp
had a few pieces of bone, a large buckle (35 × 45 mm) and several pieces of
an old leather belt in her stomach; and a large cub ate its mother after she
had been dead for over a week (Fig. 2.7).

It is important to note that many of the food items eaten by brown
hyaenas only provided a meal for one hyaena at a time. The average
number of brown hyaenas feeding per item was 1.3 compared with 3.2 for
spotted hyaenas.

The diet of the brown hyaena in the southern Kalahari is similar to its
diet in the central Kalahari (Owens & Owens 1978) and in agricultural
areas of the Transvaal, South Africa (Skinner 1976). Along the Namib
Desert coast, the brown hyaena has a far more restricted diet, consisting
predominantly of Cape fur seals (Skinner & van Aarde 1981, Siegfried
1984, Stuart & Shaughnessy 1984, Goss 1986).

2.2.2   Differences in diet between adults and cubs

The milk diet of brown hyaena cubs is supplemented by food brought to
the den by adults, and by cubs older than nine months foraging for
themselves. Although the components of the diets of adult and cub brown
hyaenas are similar, they utilize them in different proportions (Table 2.7).
Cubs tend to eat more insects (Coleoptera and Isoptera) and small
mammals, and fewer large mammals and wild fruits than do adults. The
difference in the size of mammals consumed arises because adults are more
likely to bring back a small mammal carcass than a large one to the den
(Mills & Mills 1978). (Presumably the cubs' teeth and jaws are not
sufficiently developed to deal with larger bones, and also smaller carcasses

Table 2.7 Number and percentage of scats containing food items found in more than 10 per cent of scats from adult and/or cub brown hyaenas and comparisons of the frequencies each food item occurred in the scats of adults versus cubs.

Food Item	Adult n = 143	Cub n = 240	χ^2 d.f. = 1
Coleoptera	50 (35.0%)	151 (62.9%)	26.96; $p < 0.001$
Small carnivore	44 (30.8%)	99 (41.3%)	3.77; $p < 0.05$
Gemsbok	74 (51.7%)	49 (20.4%)	38.93; $p < 0.001$
Tsama	68 (47.6%)	55 (22.9%)	23.83; $p < 0.001$
Steenbok	12 (8.4%)	70 (29.2%)	21.77; $p < 0.001$
Reptile	33 (23.1%)	43 (17.9%)	1.19; $p > 0.05$
Isoptera	12 (8.4%)	62 (25.8%)	16.39; $p < 0.001$
Hartebeest	39 (27.3%)	33 (13.8%)	9.87; $p < 0.01$
Bird	22 (15.4%)	34 (14.2%)	0.03; $p > 0.05$
Brandy bush berries	21 (14.7%)	26 (10.8%)	0.90; $p > 0.05$
Gemsbok cucumber	26 (18.2%)	10 (4.2%)	19.06; $p < 0.001$
Wildebeest	22 (15.4%)	13 (5.4%)	9.56; $p < 0.01$
Springhare	17 (11.9%)	11 (4.6%)	6.02; $p < 0.02$
Hare	15 (10.5%)	4 (1.7%)	12.98; $p < 0.001$

are easier to carry.) The higher utilization by cubs of Coleoptera, the
majority of which were Tenebrionids (Mills & Mills 1978), may be due to
the tendency for these beetles to shelter in burrows, where cubs spend
more time than adults. The cubs' higher utilization of Isoptera, most of
which were harvester termites (Mills & Mills 1978), which are diurnal in
winter, may be due to the tendency for cubs to be more diurnal in winter
than adults (section 3.1). Wild fruits were never observed to be carried
back to a den, and so could only be eaten by cubs when foraging for
themselves.

2.2.3 *Hunting* vs. *scavenging*

Scavenging, as defined here, is the consumption by an animal of a carcass
which neither it, nor others of its species, killed (Kruuk 1972). All three
species of hyaena show several adaptations to scavenging:

- Their large and powerful teeth and jaws are excellent bone-crushers.
  This enables them to consume larger bones than can any of the other
  carnivores, and to obtain nutritious marrow from long bones (Fig. 2.8).

Figure 2.8 A brown hyaena uses its powerful jaws, teeth and neck to break open the leg
bone of a wildebeest.

- They are able to digest the organic matter in bone completely and obtain the maximum energy (Kruuk 1972), which no other carnivore can do.
- They are able to cover large distances in search of carrion.

Brown hyaenas scavenged most of their food (Table 2.5), but what percentage of their diet comprised killed prey? On average a brown hyaena in the southern Kalahari consumes 2.8 kg of food (section 2.2.5), and moves 31.1 km in a 24 h period (section 3.1). From a sample of 2406 km of following adults (equivalent to 77 nights), the total biomass of animals killed was calculated as 18.6 kg. Not all this killed prey was eaten by the hyaenas which did the killing; an entire bat-eared fox and half a springbok lamb were carried back to the den for the cubs. Thus, the total biomass of prey killed and eaten by a brown hyaena in 77 nights was approximately 12.2 kg. Therefore, in 77 nights a hyaena consumes approximately 2.8×77 = 216 kg of food, of which 5.6% is killed.

In the central Kalahari, brown hyaenas also scavenge nearly all their vertebrate food (Owens & Owens 1978), as they do on the Namib Desert coast, in spite of the fact that there is a plentiful supply of Cape fur seal pups which are apparently easy to kill (Goss 1986).

Many of the scavenged food items eaten by brown hyaenas were odd legs (Fig. 2.8), skulls, and small pieces of bone, the origins of which could not be determined. I was only able to do this at some fresh carcasses (Table 2.8). Lions provided 42.7% of carcasses of known provenance, but the size of lion prey made their carcasses easier to detect than those of smaller carnivores such as cheetah. Non-violent mortality was also important, providing 29.9% of scavenged food. Additionally, these carcasses often contained larger amounts of meat than those killed by predators. Black-backed jackals and caracals occasionally provided brown hyaenas with a small, but high quality carcass. The relationships between brown hyaenas and the other carnivores are discussed in section 2.6.

2.2.4   Wild fruits

The tsama (Fig. 2.5) and gemsbok cucumber are the most important fruits in the brown hyaena's diet. Both grow from ground runners. The tsama is a melon-like fruit with an edible mass of 300–700g; the gemsbok cucumber is oval, with a spiny skin and an edible mass of 100–200g. Both fruits are eaten by many other species, particularly rodents and ungulates (Leistner 1967, personal observations), and by Bushmen (Story 1958).

Fruits become available from about March each year, although the perennial cucumbers appear slightly earlier than do the annual tsamas. Tsamas are naturally frost resistant and can remain edible for over a year. Gemsbok cucumbers, on the other hand, are susceptible to frost and the bulk of the crop has usually disappeared by July. The utilization of both

Table 2.8   Number of brown hyaenas feeding from scavenged carcasses where the agent of supply was known.

Species	Lion	Leopard	Cheetah	Spotted hyaena	Black-backed jackal	Caracal	Martial eagle	Man	Non-violent	Total
Wildebeest adult	36	0	0	2	0	0	0	1	25	64 (39.0%)
Springbok adult	1	2	2	0	0	1	0	0	5	11 (6.7%)
Springbok lamb	0	0	0	0	3	1	1	0	3	8 (4.9%)
Gemsbok adult	18	0	0	0	0	0	0	0	0	18 (11.0%)
Gemsbok calf	1	2	0	6	0	0	0	0	0	9 (5.5%)
Hartebeest adult	10	0	0	0	0	0	0	4	0	14 (8.5%)
Hartebeest calf	0	0	0	1	0	0	0	0	0	1 (0.6%)
Eland adult	4	0	1	0	0	0	0	0	6	11 (6.7%)
Steenbok	0	0	0	0	4	4	0	0	2	10 (6.1%)
Bat-eared fox	0	0	0	0	0	0	0	0	5	5 (3.0%)
Black-backed jackal	0	1	0	0	0	0	0	0	3	4 (2.4%)
African wild cat	0	0	0	0	0	3	0	0	0	3 (1.8%)
Springhare	0	0	0	0	1	2	0	0	0	3 (1.8%)
Small rodent	0	0	0	0	3	0	0	0	0	3 (1.8%)
Total	70(42.7%)	5(3.0%)	3(1.8%)	9(5.5%)	11(6.7%)	11(6.7%)	1(0.6%)	5(3.0%)	49(29.9%)	164

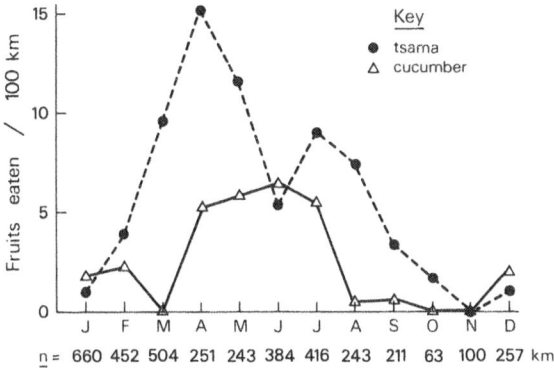

Figure 2.9 Monthly utilization of tsamas and gemsbok cucumbers by brown hyaenas expressed as number eaten per 100 km followed.

Table 2.9   Tsama and gemsbok cucumber densities in the Kwang territory, 1976-1979.

Year	Tsamas/ha	Gemsbok cucumbers/ha
1976	1293	75*
1977	5	1239
1978	0	208
1979	298	156*

* Undercounts as they were carried out in July after frost had damaged the crop

fruits by brown hyaenas through the year reflects their seasonal availability (Fig. 2.9).

Both fruits grow in the dunes (Fig. 2.10), although crop sizes vary annually (Table 2.9), due to the temporal distribution and total amount of rainfall. Although given to large fluctuations, the gemsbok cucumber crop is more reliable, albeit shorter lasting than the tsama. The fruits are patchily distributed (Fig. 2.10). Tsama patches appeared to be randomly distributed from year to year, whereas gemsbok cucumbers tended to be more concentrated, for example in the area ENE of Kwang in the brown hyaena study area (Fig. 2.10).

Both fruits have a low calorific value (30–100 kJ/100g, A. Wehmeyer personal communication), with 22 tsamas yielding equivalent energy to 1 kg of fresh meat (Mills & Mills 1978). (The most I observed a brown hyaena eat in one night was 18.) They are, however, rich in trace elements and vitamin C. In addition the fruits have a moisture content of over 90%, and are an important source of water for brown hyaenas, considering that much of their vertebrate food is dry when eaten.

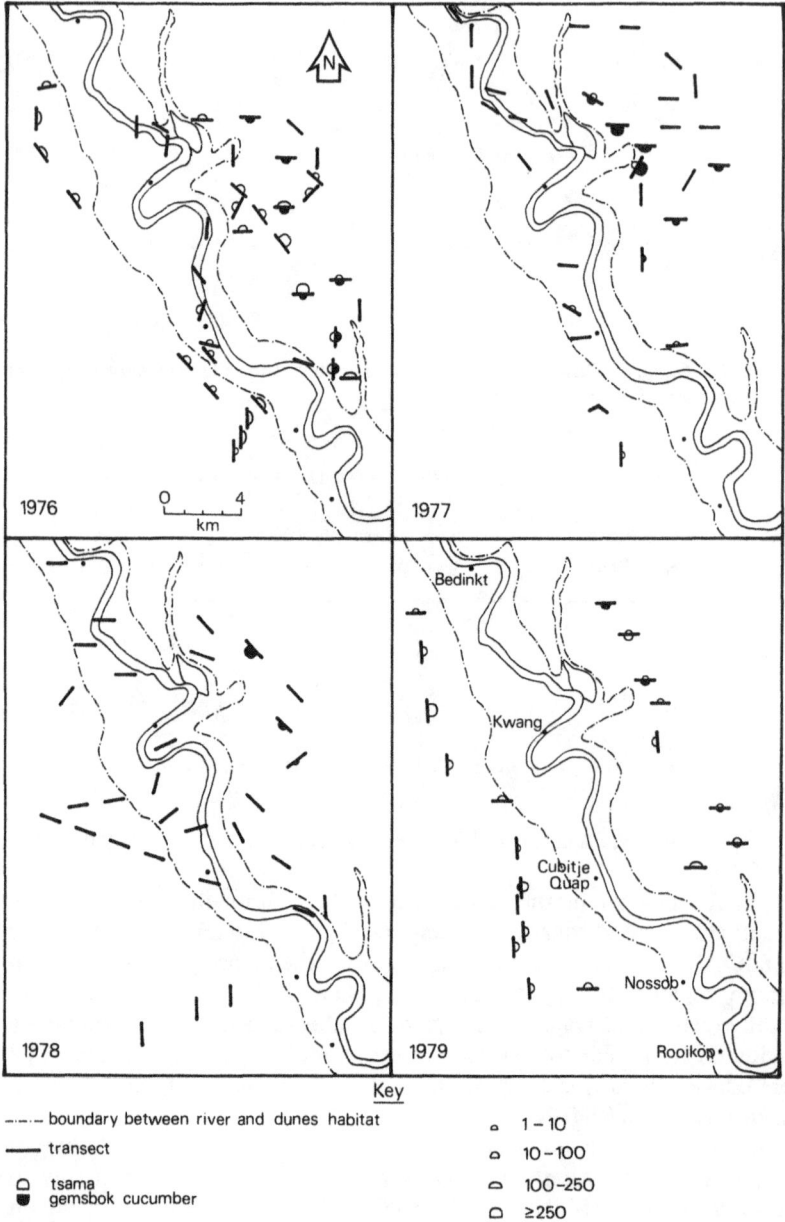

Figure 2.10 Distribution and size of tsama and gemsbok cucumber patches in the brown hyaena Kwang territory, 1976–1979.

Other fruits occasionally eaten by brown hyaenas are the highly nutritious berries of the brandy bush, the berries of merremia, an annual ground creeper, and the fungus the Kalahari truffle, which is related to the European truffle (Table 2.5).

2.2.5   Consumption

Most large carnivores live under what Schaller (1972) termed a 'feast or famine' regime. Brown hyaenas are capable of eating large quantities of food quickly, but rarely have the opportunity to do so. There is generally less fluctuation in nightly consumption rates in brown hyaenas than in those of other large carnivores. In one exceptional instance, two consumed roughly 15 kg of a springbok carcass in three hours.

I measured the average daily consumption rate of brown hyaenas in the southern Kalahari by following two lactating females for five periods of two consecutive nights, and one period of three consecutive nights. On all occasions they had been followed for part of the night prior to the initial full night, and had not fed exceptionally well. Furthermore, the mean distance travelled by the hyaenas during these 13 nights was 27.8 km, which is close to the mean distance of 31.1 km travelled per night by a brown hyaena (section 3.1). This suggests that these were typical nights of foraging.

All food items eaten were recorded and a subjective assessment of the mass of each was made using known weights from the literature for mammals (Smithers 1971), and from my own measurements for bones, wild fruits, eggs, lizards, and beetles. The results are summarized in Table 2.10. The 138 food items eaten give a total biomass of 36.48 kg and a mean of 2.8 ± 0.78 kg/hyaena/day, or 0.07 kg of food/kg of brown hyaena/day.

Table 2.10   Calculated biomass of food eaten by two lactating adult female brown hyaenas in 13 nights.

Food item	Number eaten	Biomass (kg) eaten	% of total biomass
Animal	71	18.7	51.2
Meat, skin and bone	31	16.3	44.7
Ostrich eggs	3	2.2	6.0
Unidentified small items	30	0.2	0.5
Vegetable	67	17.8	48.8
Tsama	28	15.1	41.4
Gemsbok cucumber	18	2.3	6.3
Truffle	21	0.4	1.1
Total	138	36.5	100.0

This figure is somewhat lower than the spotted hyaena's daily consumption rate in the southern Kalahari, but is close to consumption rates of spotted hyaenas in other areas (section 2.3.4).

2.3   Spotted hyaena diet

2.3.1   Overall diet

Spotted hyaenas eat almost any kind of mammal as well as birds, reptiles, amphibians, insects, molluscs, and vegetable matter, and even, on occasions, 'non-food' items such as paper and motor car tyres (Pienaar 1969, Kruuk 1972). Their habit of scavenging from rubbish dumps and around towns is legendary. However, in the southern Kalahari they are rather specialized feeders, far more so than brown hyaenas.

Table 2.11 records all food items on which spotted hyaenas were recorded to feed while being followed, or when tracking spoor. Clearly, large (over 80 kg) and medium-sized (12–80 kg) mammals are their most important food. Of 1110 feeding hyaenas, 94.1% were feeding on an animal as large, or larger than, a steenbok, the adults of which weigh about 12 kg. Ignoring the observations of spotted hyaenas chewing on bones, 35.8% feeding on mammals were doing so on gemsbok (Fig. 2.11). Next in importance were wildebeest 22.3% (Fig. 2.12), followed by springbok

Figure 2.11   Three spotted hyaenas feed on an adult gemsbok.

Table 2.11   Spotted hyaena diet from direct observations and tracking spoor.

Species	No. of carcasses from kills and scavenging	No. of hyaenas eating from kills and scavenging	No. of carcasses from kills only	No. of hyaenas eating from kills only
Mammalia	298 (86.1%)	1038 (93.5%)	105 (95.5%)	520 (97.7%)
Gemsbok adult	27 (7.8%)	141 (12.7%)	13 (11.8%)	80 (15.0%)
Gemsbok calf	45 (13.0%)	199 (17.9%)	42 (38.2%)	190 (35.7%)
Wildebeest adult	28 (8.1%)	156 (14.1%)	10 (9.1%)	69 (13.0%)
Wildebeest calf	12 (3.5%)	56 (5.0%)	10 (9.1%)	49 (9.2%)
Springbok adult	19 (5.5%)	88 (7.9%)	2 (1.8%)	14 (2.6%)
Springbok lamb	7 (2.0%)	8 (0.7%)	5 (4.5%)	6 (1.1%)
Hartebeest adult	12 (3.5%)	85 (7.7%)	3 (2.7%)	26 (4.9%)
Eland adult	9 (2.6%)	51 (4.6%)	3 (2.7%)	19 (3.6%)
Eland calf	7 (2.0%)	32 (2.9%)	7 (6.4%)	32 (6.0%)
Kudu adult	2 (0.6%)	23 (2.1%)	2 (1.8%)	23 (4.3%)
Steenbok adult	5 (1.4%)	11 (1.0%)	1 (0.9%)	1 (0.2%)
Steenbok lamb	1 (0.3%)	1 (0.1%)	1 (0.9%)	1 (0.2%)
Porcupine	3 (0.9%)	7 (0.6%)	2 (1.8%)	4 (0.8%)
Springhare	6 (1.7%)	6 (0.5%)	1 (0.9%)	1 (0.2%)
Duiker	3 (0.9%)	5 (0.5%)	0 (0.0%)	0 (0.0%)
Hare	2 (0.6%)	4 (0.4%)	2 (1.8%)	4 (0.8%)
Aardvark	1 (0.3%)	2 (0.2%)	0 (0.0%)	0 (0.0%)
Spotted hyaena	1 (0.3%)	2 (0.2%)	0 (0.0%)	0 (0.0%)
Small rodent	1 (0.3%)	1 (0.1%)	1 (0.9%)	1 (0.2%)
Skin and bones	107 (30.9%)	160 (14.4%)	0 (0.0%)	0 (0.0%)
Aves	12 (3.5%)	35 (3.2%)	4 (3.6%)	11 (2.1%)
Ostrich adult	3 (0.9%)	12 (1.1%)	1 (0.9%)	6 (1.1%)
Ostrich chick	3 (0.9%)	5 (0.5%)	3 (2.7%)	5 (0.9%)
Ostrich eggs*	6 (1.7%)	18 (1.6%)	-	-
Insects **	1 (0.3%)	1 (0.1%)	1 (0.9%)	1 (0.0%)
Fruits (tsama)**	6 (1.7%)	7 (0.6%)	-	-
Miscellaneous ***	9 (2.6%)	9 (0.8%)	-	-
Unidentified	20 (5.8%)	20 (1.8%)	0 (0.0%)	0 (0.0%)
Total	346	1110	110	532

*   Three nests and three single eggs
**   Number of patches fed on
***   Faeces, human wastage, etc.

10.1%, hartebeest 8.9%, and eland 8.7%. Large and medium-sized mammals also made up the major portion of spotted hyaenas' diets in other areas (Kruuk 1972, 1980, Bearder 1977, Smuts 1979, Tilson et al. 1980, Henschel 1986).

The few observations of spotted hyaenas eating wild fruits are in contrast to the brown hyaena's predilection for them. Although availability may have been a factor, in that tsamas and gemsbok cucumbers were few during the years of the spotted hyaena study, they passed through tsama patches without eating any fruits. Moreover, the few fruits they did eat were merely bitten open and a few small pieces of flesh ingested. However, in a sample of 149 spotted hyaena scats collected during years of abundant wild

Figure 2.12 Two spotted hyaenas feed on a wildebeest yearling they have just killed. Note the blood around their faces – an indication that the hyaenas killed.

fruit crops (1974–1976), 24 (16%) contained wild fruit pips (Mills 1978b). Stuart (1976) also recorded wild fruits as an occasional item in the spotted hyaena's diet in the Namib Desert. Because they obtain moisture from the blood of their prey, spotted hyaenas may not require the moisture from these water-storing plants to the extent that brown hyaenas do.

2.3.2   Hunting vs. scavenging

Spotted hyaenas have been regarded as predominantly scavengers by most naturalists. The first author to describe them as significant predators was Eloff (1964), from the southern Kalahari, where tracks in the sand revealed spotted hyaena kills, mainly gemsbok. It was the study of Kruuk (1966, 1970, 1972) in East Africa which showed that spotted hyaenas can be significant predators, having a major influence on many aspects of their prey's behaviour. Since then several other studies have provided evidence for spotted hyaenas killing a substantial proportion of their food (Bearder 1977, Smuts 1979, Tilson et al. 1980, Henschel 1986).

In the present study it was possible to follow hyaenas and, therefore, to quantify accurately their hunting activities. Apart from observations of hyaenas killing, I often arrived at a carcass with hyaenas already feeding on it. In these cases I was able to determine if the hyaenas were feeding on a killed or scavenged carcass, by using several clues:

- If the hyaenas had killed, their heads and forequarters would be covered in blood (Fig. 2.12).
- If a carcass had obviously been dead for some time and, therefore, scavenged, there would be little blood on it and *rigor mortis* would have set in.
- The remains of the carcass and the amount that the hyaenas had eaten (judged by the amount their stomachs were distended), indicated whether other animals had previously fed on it.
- Any doubt could usually be dispelled by examining the tracks in the sand.

In the southern Kalahari, killed prey forms the majority of the spotted hyaena's diet. If the last five small items in Table 2.11, as well as skin and bones, are excluded, then 55.3% of the carcasses they fed on were kills, and 59.3% of the hyaenas observed feeding were feeding on kills. However, these figures are biased in favour of scavenging, as the amount of meat available per carcass is generally more from kills than it is from scavenged carcasses, which have often been partially eaten by other animals before the hyaenas find them (Fig. 2.13). In addition, all things being equal, the number of hyaenas feeding from a carcass was higher if it was killed than if it was scavenged (Fig. 2.14) (1). In terms of kilograms of meat available for consumption, 73% was calculated to come from kills, and 27% from scavenging (see Table 2.15).

Figure 2.13   A sub-adult male spotted hyaena scavenges from an eland carcass.

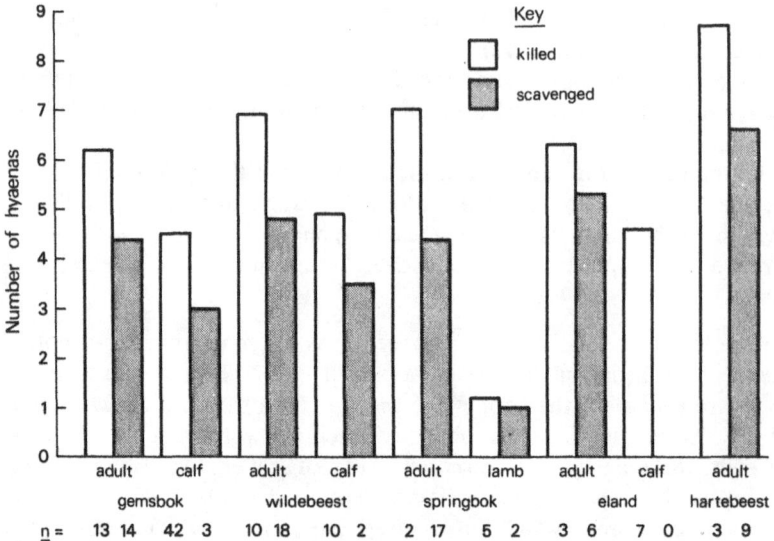

Figure 2.14 Mean number of spotted hyaenas which fed on different species which were killed or scavenged.

How do these figures compare with those from other areas? Kruuk (1972) recorded that 43.1% of food items and 54.9% of hyaenas observed feeding in the Serengeti, and 82.2% of food items and 90.6% of hyaenas in the Ngorongoro Crater, were feeding on their own kills, which comprised mainly zebra and wildebeest. Henschel (1986) studied a spotted hyaena clan in the Kruger National Park, where conditions for direct observations were more difficult than in the Kalahari or in East Africa. Cause of death could not be established for 65.0% of ungulate carcasses on which spotted hyaenas were observed feeding, but at least 24 (16.2%), as well as several non-ungulate species, were killed by the hyaenas. He concluded that these spotted hyaenas killed less, and had a less specialized diet, than those in the Kalahari and East Africa.

Table 2.12 shows that in the southern Kalahari spotted hyaenas are more likely to kill the young of gemsbok, wildebeest, springbok, and eland than to scavenge them. Hartebeest calves, on the other hand, are rarely killed by spotted hyaenas. Gemsbok, wildebeest, and eland adults (here defined as animals over one year of age), are killed and scavenged in equal proportions, whereas hartebeest and springbok adults are most likely to be scavenged. The reasons for these discrepancies and the mechanisms involved will be discussed in section 2.3.3.

Most of the carcasses scavenged by spotted hyaenas resulted from non-violent mortality, followed by lion and cheetah kills (Table 2.13). This is similar in the case of brown hyaenas, except that the latter also

Table 2.12   The number of carcasses of the more important food species that were either killed or scavenged by spotted hyaenas from all observations; i.e. following animals and opportunistic sightings.

Species	Killed	Scavenged
Gemsbok adult	14 (50.0%)	14 (50.0%)
Gemsbok calf	52 (92.9%)	4 (7.1%)
Wildebeest adult	18 (50.0%)	18 (50.0%)
Wildebeest calf	16 (88.9%)	2 (11.1%)
Springbok adult	2 (7.7%)	24 (92.3%)
Springbok lamb	8 (80.0%)	2 (20.0%)
Hartebeest adult	3 (33.3%)	9 (66.6%)
Eland adult	4 (44.4%)	5 (55.5%)
Eland calf	7 (100.0%)	0 (0.0%)
Ostrich	3 (60.0%)	2 (40.0%)
Total	127 (61.4%)	80 (38.6%)

Table 2.13   Number of spotted hyaenas feeding from scavenged carcasses, where the agent of supply was known.

Species	Lion	Leopard	Cheetah	Caracal	Black-backed jackal	Non-violent	Total
Springbok adult	0	5	34	3	0	31	73 (23.0%)
Springbok lamb	0	0	3	0	0	1	4 (1.3%)
Wildebeest adult	25	0	0	0	0	40	65 (20.4%)
Wildebeest calf	0	0	0	0	0	7	7 (2.2%)
Gemsbok adult	12	0	0	0	0	46	58 (18.2%)
Gemsbok calf	3	4	0	0	0	5	12 (3.8%)
Hartebeest adult	5	0	0	0	0	52	57 (17.9%)
Eland adult	9	0	0	0	0	10	19 (6.0%)
Steenbok	0	0	8	0	0	2	10 (3.1%)
Ostrich	0	4	0	0	0	2	6 (1.9%)
Springhare	0	0	0	1	1	0	2 (0.6%)
Aardvark	0	0	0	0	0	2	2 (0.6%)
Spotted hyaena	2	0	0	0	0	0	2 (0.6%)
Duiker	0	0	1	0	0	0	1 (0.3%)
Total	56(17.6%)	13(4.1%)	46(14.5%)	4(1.3%)	1(0.3%)	198(62.3%)	318

scavenged more frequently from smaller carnivores (Table 2.8). Most of the carcasses scavenged by spotted hyaenas from predators were done so after the predators had abandoned them. Interactions between spotted hyaenas and the other carnivores are described in section 2.6.

2.3.3   Prey selection

What sort of animals are killed by spotted hyaenas? Do they take some species, cohorts, or individuals in preference to others?

Table 2.14   Kills made by spotted hyaenas on nights that they were being followed.

Species	Age-class	Number killed
Gemsbok	Adult	5 (5.2%)
	Sub-adult	5 (5.2%)
	Calf	42 (43.3%)
	Total	52 (53.6%)
Wildebeest	Adult	2 (2.1%)
	Sub-adult	3 (3.1%)
	Calf	10 (10.3%)
	Total	15 (15.5%)
Eland	Adult	1 (1.0%)
	Calf	6 (6.2%)
	Total	7 (7.2%)
Springbok	Adult	1 (1.0%)
	Sub-adult	1 (1.0%)
	Lamb	5 (5.2%)
	Total	7 (7.2%)
Hartebeest	Adult	2 (2.1%)
	Sub-adult	1 (1.0%)
	Total	3 (3.1%)
Ostrich	Adult	1 (1.0%)
	Chick	3 (3.1%)
	Total	4 (4.1%)
Kudu	Adult	2 (2.1%)
Steenbok	Adult	1 (1.0%)
	Lamb	1 (1.0%)
	Total	2 (2.1%)
Hare	Adult	1 (1.0%)
	Leveret	1 (1.0%)
	Total	2 (2.1%)
Springhare	Adult	1 (1.0%)
Porcupine	Adult	1 (1.0%)
Mouse	Adult	1 (1.0%)
Total	Adult	19 (19.6%)
	Sub-adult	10 (10.3%)
	Juvenile	68 (70.1%)

Table 2.14 lists only those animals that were killed when hyaenas were being followed, to eliminate the possible bias towards larger animals. Clearly juveniles are important prey, making up 70.1% of their kills.

Table 2.15 presents all spotted hyaena kills documented during the study. Wherever possible the prey was sexed and aged according to the state of eruption or wear of the mandibular teeth (see Appendix C).

2.3.3.1 GEMSBOK

Gemsbok calves are the most important prey for southern Kalahari spotted hyaenas (Fig. 2.15), forming 43.3% of the kills in Table 2.14. Counts of gemsbok gave an average figure of 15.8% calves in the population (Table 2.3), whereas 80.8% of gemsbok killed by spotted hyaenas were calves (Table 2.14), showing a strong selection for calves (2). These included

Figure 2.15 A spotted hyaena kills a young gemsbok calf after a chase of 1.4 km. Note the bulging eye of the calf where the hyaena had bitten it in the head.

calves at all stages of development up to one year old, age-class 1 (0–6 months), and age-class 2 (7–12 months), calves being represented in equal proportions (Table 2.15). Assuming that spotted hyaenas make a kill once every 2.8 days (section 2.3.4), that 43.3% of these kills are gemsbok calves (Table 2.14), that 4.5 spotted hyaenas feed off a gemsbok calf kill (Fig. 2.14), and that there are 100 spotted hyaenas in the study area (section 1.4.3), then in one year spotted hyaenas will remove 1254 gemsbok calves from the population. The impact that this may have on the population is discussed in section 2.5.

Because of the difficulty in separating sub-adult (age-class 3) gemsbok from adults (age-class 4 and above) in the field, I was unable to quantify the relative number of sub-adults in the population. However, a gemsbok can only spend about 13 months in age-class 3 (Appendix C), compared to several years in age-class 4 and above, so there are obviously many more gemsbok adults than sub-adults in the population. Age-class 3 animals constituted 9% of the gemsbok killed by spotted hyaenas compared to the 12% of adult gemsbok (Table 2.15). It would appear, therefore, that spotted hyaenas also kill sub-adult gemsbok in preference to adults.

The low frequency with which adult gemsbok were killed by spotted hyaenas (Tables 2.14 and 2.15), and the fact that they were the most frequently contacted potential prey (see section 3.6.2), indicates a strong resistance by adult gemsbok to spotted hyaena predation. The small

Table 2.15   Age-classes (1-9*) of all known spotted hyaena kills.

Species	Juvenile			Sub-adult	Adult								Total
	1	2	1+2	3	4	5	6	7	8	9	Unknown	All	
Gemsbok	26	26	52	6	2	1	0	2	0	1	2	8	66 (47.5%)
Wildebeest	5	11	16	3	0	6	1	1	4	0	3	15	34 (24.5%)
Eland	6	1	7	0	3	1	0	0	0	0	0	4	11 (7.9%)
Springbok	8	0	8	1	0	1	0	0	0	0	0	1	10 (7.2%)
Ostrich	0	2	2	1	-	0	0	0	-	-	2	2	5 (3.6%)
Hartebeest	0	0	0	1	0	0	1	0	0	0	1	2	3 (2.2%)
Porcupine	0	0	0	0	-	0	-	-	-	-	3	3	3 (2.2%)
Kudu	0	0	0	0	0	0	0	0	0	0	2	2	2 (1.4%)
Steenbok	1	0	1	0	-	-	-	-	-	-	1	1	2 (1.4%)
Hare	1	0	1	0	-	-	-	-	-	-	1	1	2 (1.4%)
Springhare	0	0	0	0	-	-	-	-	-	-	1	1	1 (0.7%)
Total	47 (33.8%)	40 (28.8%)	87 (62.6%)	12 (8.6%)	5	9	2	3	4	1	16	40 (28.8%)	139

*See Appendix C

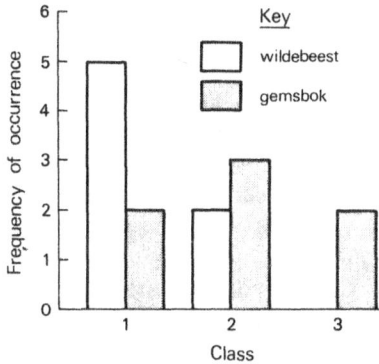

Figure 2.16 Frequency of occurrence of the categories of bone marrow from the femurs of adult gemsbok and wildebeest killed by spotted hyaenas. Class 1 has high fat content, 60–100%; class 2 moderate, 40–60%; and class 3 low, < 40%.

sample of kills in Table 2.15 does not point to any age-class selection for adult gemsbok and only one of the gemsbok killed appeared to be handicapped – a sub-adult with one horn twisted across its face. However, the amount of fat in the bone marrow of the femur of an ungulate gives an indication of its condition, since the fat stored in the marrow of the leg bones is usually the last to be utilized by an animal suffering from malnutrition (Cheatum 1949, Schaller 1972, Brooks *et al.* 1977). Visual inspection of the marrow of seven adult gemsbok caught by spotted hyaenas suggested that three had severe and two had moderate fat depletion (Fig. 2.16). These at least were probably in poor condition.

In the Namib Desert, Tilson *et al.* (1980) found that spotted hyaenas regularly killed adult gemsbok, with evidence that they selected older individuals. They point out that they were unable to follow hyaenas, or to track their spoor on a regular basis, so that many of the carcasses they found had been dead for some time. Not only did this bias their results towards adult animals, because calves would be consumed far more quickly and without leaving such obvious traces, but also it often made it impossible to establish the cause of death.

2.3.3.2 WILDEBEEST
The second most important prey species for spotted hyaenas in the southern Kalahari is wildebeest (Tables 2.14 and 2.15). Calves made up a smaller proportion of kills of wildebeest than of gemsbok. Nevertheless, 66.7% of wildebeest killed were calves (Table 2.14), compared with 18.6% calves in the population (Table 2.3) – another strong selection for calves (3). In contrast to the Serengeti and Ngorongoro, where the majority of wildebeest calves killed by spotted hyaenas were very young (Kruuk 1972), there was no selection for young calves in the southern Kalahari.

Sub-adult wildebeest were often also caught disproportionately (4) – 20% in the killed sample as opposed to 11.0% in the population (Table 2.3).

Adult wildebeest were caught less often than their frequency in the population (5), of which they comprised 70.4% (Table 2.3). No trend in the age distribution of the small sample of adult kills is apparent (Table 2.15). The sex ratio was eight males and seven females, which shows a tendency for males to be killed (6), as the sex ratio in the population was 1:1.98 (Table 2.3). The condition of the adult wildebeest killed generally appeared to be good (Fig. 2.16), although one kill was an adult female with severe sarcoptic mange. Because of their relative availability, wildebeest are not so important as prey animals for southern Kalahari spotted hyaenas (15.5% of kills) as they are for hyaenas in the Serengeti (53.3% of kills) and Ngorongoro (72.9% of kills) (Kruuk 1972).

2.3.3.3 ELAND
Eland calves were selected for and adults were almost immune from spotted hyaena predation (Tables 2.14 and 2.15). As will be discussed in section 3.6.5, there appeared to be a selection for unfit, or otherwise vulnerable adults, but quantitative data are lacking.

2.3.3.4 SPRINGBOK
For the short period that young springbok lambs are available they are preyed upon by spotted hyaenas (Tables 2.14 and 2.15). The frequency with which springbok lambs were caught may be under-represented in the data, as they were usually hunted solitarily and consumed rapidly. Thus, for example, if in one night four hyaenas from a clan each caught a lamb, I would probably only have recorded the kill of the one I was following.

Springbok adults, in spite of their abundance, are rarely caught by spotted hyaenas (Tables 2.14 and 2.15).

2.3.3.5 OTHER SPECIES
Two other abundant large animals in the southern Kalahari are hartebeest, whose numbers fluctuated widely in the study area (Fig. 2.1, Appendix B), and ostrich, which were more evenly distributed (Van der Walt et al. 1984). Neither of these species was hunted to any extent by spotted hyaenas. One of the adult hartebeest killed had severe sarcoptic mange, and of the two adult ostriches killed, one was in a weak state and the other was caught on its nest.

Smaller, common animals, such as steenbok, springhare, hare, and porcupine were usually ignored by spotted hyaenas.

2.3.3.6 CONCLUSIONS
Spotted hyaenas are selective hunters: not only do they hunt certain species in preference to others in the same size range, but within a species

certain cohorts are selected. The species hunted are not selected solely on their numerical availability, but on the hyaenas' hunting ability and the prey's anti-predator behaviour. This is discussed in section 3.6.

With gemsbok and wildebeest the main selective criterion appears to be size; calves and sub-adults are killed in preference to adults. With adults the criteria were unclear. Apparently fit individuals were sometimes killed, although most gemsbok appeared to be in poor condition. Wildebeest bulls may have been more susceptible to spotted hyaena predation than cows. They were more prone to non-violent mortality (section 2.3.1) and, therefore, more likely to be in poor condition. Alternatively, differences in behaviour between bulls and cows may make the former more vulnerable. Small springbok lambs are vulnerable to spotted hyaena predation and so are eland calves. For the rest, opportunism is probably important. Hyaenas may be able to discern irregularities in their prey's behaviour and to capitalize on any mistakes.

2.3.4   Consumption and food availability

Kruuk (1972) recorded that a spotted hyaena is able to consume one-third of its body mass in a single meal. Several times during the present study I was struck by the volume of food spotted hyaenas were able to eat. For example, in five nights a female consumed an entire springbok lamb and a springhare, shared a complete adult springbok carcass with four others, half a springbok carcass with five others, and a complete gemsbok carcass with six others, and also chewed on some bones, for an estimated total consumption of 38.9 kg, or 7.8 kg/day. At the other extreme I followed four hyaenas for three nights, during which time they covered a distance of 118 km, and ate only three old bones. But what was the average consumption rate of spotted hyaenas?

The total biomass of food estimated to have been available to spotted hyaenas during 7160 km of observations was 8076 kg (Table 2.16). On average a spotted hyaena moves 27.1 km/24 h period (section 3.1). Therefore, from kills:

95 kills were made in 7160 km = 1 kill/75.4 km = 1 kill/2.8 days.
5864 kg were available to 470 hyaenas (Table 2.16).
Therefore, one hyaena has available to consume 12.5 kg/2.8 days = 4.5 kg/day.

From scavenging:

50 carcasses were found in 7160 km = 1 carcass/143.2 km = 1 carcass/5.3 days.
2212 kg were available to 240 hyaenas (Table 2.16).
Therefore, one hyaena has available to consume 9.2 kg/5.3 days = 1.7 kg/day.

Table 2.16 Biomass of food available to spotted hyaenas from kills and scavenged carcasses from following hyaenas for 7160km. The average mass for each type of carcass was calculated from figures provided by Smithers (1971), Meissner (1982) and personal observations. The percentage wastage (horns, large bones, stomach contents) from killed carcasses was 33% for large animals (over 80kg), 10% for medium-sized and small animals (5-80kg) and 0% for very small animals (0-5kg), and were subtracted from each carcass. The amount subtracted from scavenged carcasses depended on the condition of the carcass when it was first eaten by the hyaenas.

Species	Mass (kg)	Kills				Scavenged			
		No. of carcasses	No. of hyaenas feeding	Biomass eaten (kg)	kg eaten per hyaena	No. of carcasses	No. of hyaenas feeding	Biomass	kg eaten eaten (kg)
Gemsbok adult	200.0	10	61	1340.0	22.0	9	48	909.0	18.9
Gemsbok calf	50.0	42	190	1890.0	9.9	3	10	51.0	5.1
Wildebeest adult	200.0	5	31	670.0	21.6	5	45	396.0	8.8
Wildebeest calf	50.0	10	57	450.0	7.9	1	4	17.0	4.3
Eland adult	500.0	1	6	335.0	55.8	2	10	340.0	34.0
Eland calf	80.0	6	29	432.0	14.9	0	0	0.0	0.0
Springbok adult	40.0	2	14	72.0	5.1	12	59	209.0	3.5
Springbok lamb	5.0	5	6	25.0	4.2	1	1	3.0	3.0
Hartebeest adult	145.0	3	26	291.0	11.2	6	45	130.0	2.9
Ostrich adult	100.0	1	3	67.0	22.3	2	6	100.0	16.7
Ostrich chick	2.5	3	5	7.5	1.5	0	0	0.0	0.0
Kudu adult	200.0	2	33	268.0	8.1	0	0	0.0	0.0
Steenbok adult	10.0	0	0	0.0	0.0	2	3	13.0	4.5
Steenbok lamb	2.5	1	1	2.5	2.5	0	0	0.0	0.0
Porcupine adult	10.0	1	3	9.0	3.0	0	0	0.0	0.0
Hare adult	1.5	1	3	1.5	0.5	0	0	0.0	0.0
Hare leveret	0.5	1	1	0.5	0.5	0	0	0.0	0.0
Springhare adult	3.0	1	1	3.0	3.0	5	5	10.0	2.0
Aardvark adult	50.0	0	0	0.0	0.0	1	2	17.0	8.4
Spotted hyaena cub	25.0	0	0	0.0	0.0	1	2	17.0	8.4
Total		95	470	5864.0		50	240	2212.0	

On average, a hyaena had available 4.5 kg/day from kills + 1.7 kg/day from scavenging = 6.2 kg/day.

Assuming that the average spotted hyaena weighs 61.6 kg (the weighted mean for the composition of a 'typical' southern Kalahari clan), this gives 0.10 kg of food available/kg of hyaena/day. (This figure does not include the pieces of bone available and consumed by spotted hyaenas.) I have called this food 'available to consume' because the hyaenas did not always consume the carcasses as completely as possible, but the left-overs were not measured.

The 'available for consumption' figure for southern Kalahari spotted hyaenas, is considerably higher than the consumption figures for spotted hyaenas from other areas (Table 2.17). There are several possible reasons for this:

- The large distances travelled by southern Kalahari spotted hyaenas demanded a higher food intake.
- Smaller numbers of hyaenas fed from same-sized carcasses, compared, for example, with their East African counterparts. This meant that there was more food available per hyaena per carcass.
- Southern Kalahari spotted hyaenas tended to under-utilize carcasses, possibly because having often travelled far from the den, females had to get back to their cubs and could not afford to wait and digest the food before having a second feed on the carcass. The relationship between food availability and density of hyaenas is discussed in section 4.6.

Table 2.17   Daily consumption rates of spotted hyaenas from various areas.

Area	kg/day	kg/kg of hyaena/day	Source
Kalahari	6.2	0.10	This study
Ngorongoro Crater			
Wildebeest calving period	5.4	0.10	Kruuk 1972
Dry season	2.0	0.04	
Namib Desert	4.0	-	Henschel & Tilson 1988
Kruger National Park	3.8	0.06	Henschel 1986
Umfolozi Game Reserve	3.8	0.07	Green et al. 1984
Serengeti	3.0	0.06	Kruuk 1972

2.4   Diets of the other large carnivores and ecological separation of the predators

Three large carnivores, lion, leopard, and cheetah, are potential competitors for prey with spotted hyaenas in particular, and potential providers of food for both hyaena species. Most of the data in this section were obtained from kills fortuitously located, rather than from actually following the predators, as was the case with spotted hyaenas. Therefore, there is a danger of bias against smaller animals. This is discussed in Mills (1984b), and is referred to here where necessary.

2.4.1   Lion

Table 2.18 records lion kills. In addition, lions scavenged 11 wildebeest, three eland, two springbok, one hartebeest, and one gemsbok, i.e. 4.6% of all carcasses on which they fed. Wildebeest (Fig. 2.17) and gemsbok are clearly the dominant food of lions, contributing 69.2% of all carcasses on which they fed, the majority of which were kills. Most lion kills were adults (75.1%). Species from which juveniles were, relatively, more often killed, were gemsbok (26.7% of gemsbok kills), and springbok (22.9% of springbok kills) (Table 2.18).

Lions did not appear to select for sex in adult wildebeest (Table 2.19), although they did for older individuals (Fig. 2.18) (7). The age structure of wildebeest dying non-violently was similar to that of those killed by lions (Fig. 2.18), although in this case there was no difference from the age structure of the survivors (8). The high incidence of non-violent wildebeest mortality during the drought years has been mentioned in section 2.2.1. During the period 1973–1977, 70 wildebeest killed by lions were found in the spotted hyaena study area, compared with five which had died

Table 2.18   An analysis of lion kills from the southern Kalahari.

Species	Adult	Sub-adult	Juvenile	Total
Wildebeest	106	12	19	137 (37.0%)
Gemsbok	78	10	32	120 (32.4%)
Springbok	35	2	11	48 (13.0%)
Hartebeest	23	2	1	26 (7.0%)
Eland	14	1	1	16 (4.3%)
Ostrich	14	1	0	15 (4.1%)
Porcupine	7	0	0	7 (1.9%)
Aardvark	1	0	0	1 (0.3%)
Total	278 (75.1%)	28 (5.7%)	64 (17.3%)	370 (100.0%)

Figure 2.17 An adult male lion guards his adult wildebeest kill.

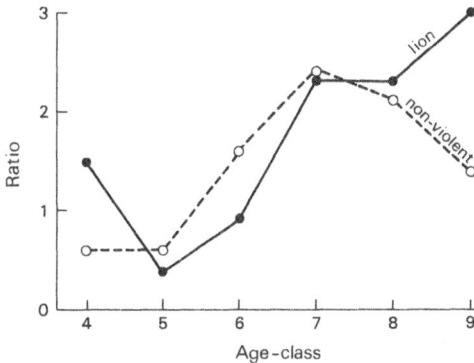

Figure 2.18 The ratio of the percentage of adult wildebeest of each age-class killed by lions ($n = 86$) to the percentage of wildebeest of each age-class randomly shot ($n = 31$), and the ratio of the percentage of adult wildebeest of each age-class which died non-violently ($n = 96$) to the percentage of each age-class in the shot sample.

non-violently. During the period 1978–1982, 67 lion-killed wildebeest carcasses were found and 109 non-violent ones (9).

More gemsbok bulls were found killed by lions than would be expected (Table 2.19), and a tendency towards higher lion predation rates in older gemsbok was found (Fig. 2.19) (10). The non-violent mortality trend is similar to that of lion predation, although the sample is too small for

Table 2.19   Sex ratios (males:females) of adult wildebeest, gemsbok and hartebeest in the living population, and from lion kills and non-violent mortality.

Species	Sex ratio in population	Sex ratio in kills	χ^2 Population vs. kills	Sex ratio of non-violent deaths	χ^2 Population vs. non-violent
Wildebeest	1:1.98	1:1.4 $n=89$	2.46 d.f.=1; $p>0.05$	1:0.8 $n=103$	20.95 d.f.=1; $p<0.001$
Gemsbok	1:1.19	1:06 $n=72$	8.06 d.f.=1; $p<0.01$	1:0.7 $n=27$	3.75 d.f.=1; $p>0.05$
Hartebeest	1:2.03	1:0.3 $n=19$	19.73 d.f.=1; $p<0.001$	-	-

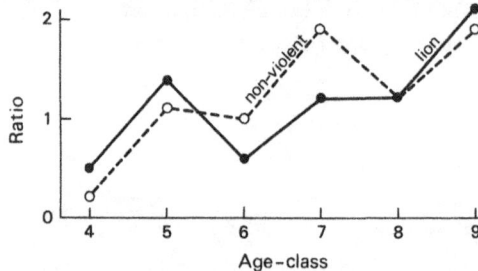

Figure 2.19   The ratio of the percentage of adult gemsbok of each age-class killed by lions (n = 45) to the percentage of gemsbok of each age-class randomly shot (n = 18), and the ratio of the percentage of adult gemsbok of each age-class which died non-violently (n = 24) to the percentage of each age-class in the shot sample.

statistical analysis. Non-violent mortality accounted for 32.8% of the known causes of adult gemsbok mortality (Fig. 2.3).

Relative to their abundance (Fig. 2.1), lions appear not to select springbok, which accounted for only 13.0% of their kills (Table 2.18). However, because springbok lambs can be consumed by a lion in less than five minutes they are doubtless under-represented in the sample. There was a selection for adult males (Table 2.20), but the sample from lion-killed springbok adults was too small for an analysis of age selection to be done.

Hartebeest constituted only 7.0% of lion kills (Table 2.18), with a selection for bulls (Table 2.19).

Of the 14 adult eland caught by lions, eight were males and six were females, and of the 14 ostriches there were seven of each sex. The

Table 2.20   Sex ratios (males:females) of adult springbok in the population, and from predator kills and non-violent mortality.

	n	Sex ratio	χ^{2*}
Population	33 164	1:1.70	-
Lion	35	1:1.06	9.91; d.f.=1; $p < 0.01$
Leopard	26	1:1.06	5.85; d.f.=1; $p < 0.02$
Cheetah	131	1:1.06	35.50; d.f.=1; $p < 0.001$
Non-violent	32	1:1.10	1.2; d.f.=1; $p > 0.05$

* Population vs. agent of mortality

relatively high incidence of porcupines in the sample of lion kills, in spite of the bias against small food items with the techniques used, is noteworthy. The phenomenon of Kalahari lions killing this and other small prey has been discussed by Eloff (1973a, b, 1984).

2.4.2   Cheetah

Table 2.21 lists the known cheetah kills. Here, small animals are significantly under-represented (Mills 1984b), and the data were, with the exception of one ostrich, one gemsbok calf, and one steenbok, collected

Table 2.21   An analysis of cheetah kills from the southern Kalahari.

Species	Number of kills
Springbok	199 (86.9%)
Adult male	84 (36.7%)
Adult female	47 (20.5%)
Adult of unknown sex	25 (10.9%)
Sub-adult	2 (0.9%)
Lamb	41 (17.9%)
Ostrich	8 (3.5%)
Wildebeest calf	8 (3.5%)
Gemsbok	6 (2.6%)
Sub-adult	2 (0.9%)
Calf	4 (1.7%)
Hartebeest calf	4 (1.7%)
Steenbok	2 (0.9%)
Bat-eared fox	2 (0.9%)
Total	229

Figure 2.20   A cheetah with its adult male springbok kill.

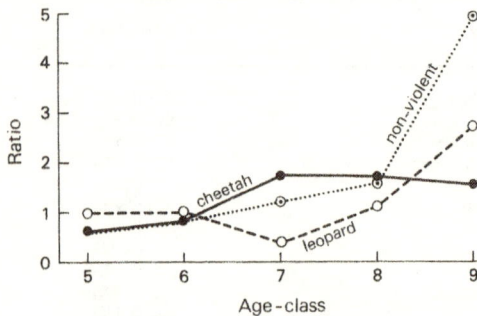

Figure 2.21 The ratio of the percentage of adult springbok of each age-class killed by cheetahs ($n = 82$) to the percentage of each age-class randomly shot ($n = 191$), the ratio of the percentage of adult springbok of each age-class killed by leopards ($n = 18$) to the percentage of each age-class in the shot sample, and the ratio of adult springbok which died non-violently ($n = 29$) to the percentage of each age-class in the shot sample.

along the river beds. Nevertheless, in this habitat springbok are clearly the cheetah's most important prey species (Fig. 2.20), accounting for 86.9% of their kills. Labuschagne (1979) found springbok comprised 41% of kills in the southern Kalahari, although his data, discerned mainly from tracking spoor, were biased towards the dunes.

Springbok lambs comprised 20.6% of all springbok killed by cheetahs. Of these 81% were under three months of age, the majority being less than four weeks old. Amongst adults there was a significant tendency for cheetahs to take males (Table 2.20), and older springbok were more prone to cheetah predation than were younger adults (Fig. 2.21) (11).

2.4.3 Leopard

Along the river beds springbok were the predominant prey for leopards (Fig. 2.22), but in the dunes leopards are more catholic in their diet (Table 2.22) (see also Bothma & le Riche (1984)).

The sex ratio of adult springbok killed by leopards also deviated significantly from the expected ratio (Table 2.20). There was again a tendency for older animals to be caught (Fig. 2.21) (12). Leopards also appeared to kill more sub-adult springbok than did cheetahs (13) (see Tables 2.21 and 2.22).

Non-violent mortality accounted for only 14.3% of all adult springbok mortality, the sex ratio of adult springbok which died non-violently did not

Figure 2.22 A leopard with its adult female springbok kill which it has dragged into a tree.

Table 2.22   An analysis of leopard kills from the southern Kalahari.

Species	River	Dunes	Total
Springbok	48	4	52 (65.0%)
Adult male	16	0	16 (20.0%)
Adult female	10	0	10 (12.5%)
Adult of unknown sex	8	1	9 (11.3%)
Sub-adult	4	1	5 (6.3%)
Lamb	10	2	12 (15.0%)
Hartebeest calf	2	3	5 (6.3%)
Steenbok	0	5	5 (6.3%)
Black-backed jackal	2	1	3 (3.8%)
Gemsbok calf	0	2	2 (2.5%)
Ostrich	1	1	2 (2.5%)
Duiker	0	2	2 (2.5%)
Wildebeest calf	1	0	1 (1.2%)
Bat-eared fox	1	0	1 (1.2%)
Cape fox	1	0	1 (1.2%)
Aardwolf	0	1	1 (1.2%)
Wild cat	0	1	1 (1.2%)
Ground squirrel	1	0	1 (1.2%)
Cheetah	1	0	1 (1.2%)
Aardvark	0	1	1 (1.2%)
Small rodent	1	0	1 (1.2%)
Total	59	21	80

differ significantly from the expected ratio (Table 2.20), and older adult springbok were more likely to succumb to non-violent mortality than were younger ones (Fig. 2.21) (14).

2.4.4   Ecological separation of the predators

The large predators in the southern Kalahari can be divided into two groups on the basis of prey selection. Lions and spotted hyaenas prey predominantly on the large to medium-sized mammals, whereas leopards and cheetahs do so on the medium-sized to small mammals. Because of their limited hunting activities, brown hyaenas are of no concern here.

The extent to which the diets of lions and spotted hyaenas overlap is lessened by differences in selection of their two most common prey species, gemsbok and wildebeest. Lions kill mainly adults and sub-adults of both species (Table 2.18), whereas spotted hyaenas kill mainly calves, particularly gemsbok calves (Table 2.15). Additionally, spotted hyaenas scavenge more food than lions. When scavenging from large and medium-sized mammals, spotted hyaenas are in direct competition with brown hyaenas (see section 8.1.1).

As mentioned in section 2.3.3.1, in the Namib Desert there is evidence that spotted hyaenas regularly kill adult gemsbok, particularly older ones

(Tilson *et al.* 1980). These are the gemsbok that also appear to be selected by lions in the southern Kalahari (Fig. 2.19). There are no lions in the Namib Desert, so it may be that in the southern Kalahari lions deprive spotted hyaenas of an important food source, by cropping adult gemsbok slightly earlier than spotted hyaenas can.

There are also several mechanisms which may reduce the amount of competition between leopards and cheetahs:

• The proportions in which they select from different age-segments of the springbok population differ slightly.
• Leopards have a wider diet than do cheetahs, particularly in the dunes.
• Cheetahs are mainly diurnal hunters and leopards are almost exclusively nocturnal, thus reducing the amount of interference competition.

2.5 The impact of predation on the prey populations

In order to answer the complex ecological question of the impact predators have on their prey, accurate data on several aspects of the dynamics of predator and prey are required. What are the numbers of predators and prey? What is the structure of the prey population? How often do the predators kill? How do they select for sex, age, and condition? What is the contribution of non-violent mortality to the dynamics of the prey populations? Much of this information, particularly the numbers of prey, is not available for the southern Kalahari (nor indeed for any other large predator–prey system in Africa). Nevertheless, there is enough information available at present to address this question qualitatively.

Unlike the nomadic ungulate populations, the predators in the southern Kalahari are largely restricted to large territories. The lion population in the Kalahari Gemsbok National Park is estimated to be between 110 and 180 (Mills *et al.* 1978) and the spotted hyaena population between 75 and 100 (section 1.4.3). The numbers of leopards and cheetahs are unknown, but there are probably about 100 leopards and 60 cheetahs. The numbers of all these predators appear to be mainly limited by the nomadic nature of their prey and their own sedentary behaviour. This results in there being periods of low food availability in each territory or home range.

Wildebeest are important prey species for both lions and spotted hyaenas (Tables 2.15 and 2.18). Lions are the most important predators of the adults and they tend to take mainly older animals (Fig. 2.18). The significant increase in non-violent mortality of wildebeest after 1977 at the end of a high rainfall period, coupled with the drop in wildebeest numbers over the succeeding years (Mills & Retief 1984a, Table 2.1), suggest that predators are of little consequence in regulating wildebeest numbers. Furthermore, the distorted sex ratio of adult wildebeest seems to be caused through a high non-violent mortality in bulls, and not due to differential predation on bulls and cows by lions. Although wildebeest calves are taken by all four predators, there is no suggestion, even in relative terms, of a heavy predation on wildebeest calves as has, for example, been found in the Ngorongoro Crater (Kruuk 1972).

Gemsbok tend to be more sedentary than the other ungulates, and no high incidences of non-violent mortality amongst them were recorded. Predation on adult gemsbok, which is mainly by lions (Table 2.18), tends to mirror non-violent mortality (Fig. 2.19), suggesting that lions are mainly removing surplus animals. However, predation on gemsbok calves is heavier than on wildebeest calves, particularly by spotted hyaenas (Tables 2.15, 2.18, 2.21, 2.22). Gemsbok are less seasonal breeders than are the other ungulates (Fig. 2.2), so that predation on calves is more evenly spread throughout the year, having a potentially heavier impact than on a synchronized breeder. The true impact of predators on gemsbok calves can only be assessed once there are accurate figures available for the number of gemsbok. It may be significant that during an aerial survey in July 1976, 1105 gemsbok were counted in the predator-free Mier Settlement on the southwest boundary of the Kalahari Gemsbok National Park, whereas only 324 were counted in an equal-sized adjacent area of similar habitat in the Park.

Cheetahs prey heavily on young springbok lambs, which have at least eight other predators from lions to martial eagles (personal observations). In the Auob river bed particularly, which has the highest density of cheetahs in the southern Kalahari, predation on young springbok lambs may be relatively heavy. However, once past the critical first few weeks of life, they become more difficult to catch.

Non-violent mortality did not appear to be heavy on adult springbok. Cheetahs and leopards (Fig. 2.21), and to a lesser extent lions, catch mainly older adults, particularly males (Table 2.20), which are the most expendable segment of the population. If predation does not exert an influence on springbok numbers it may be responsible for the distorted sex ratio of adults.

The low incidence of eland, hartebeest, and ostriches in the diets of the carnivores (Tables 2.15, 2.18, 2.21, 2.22) means that predation has an insignificant impact on these populations.

Predation, therefore, appears to have a small impact on prey numbers in the southern Kalahari, as is also the case in other self-contained ecosystems (Schaller 1972). However, predation may, on an evolutionary time-scale, have had a considerable influence on the form and behaviour of prey (Bertram 1979). The alertness of the hartebeest; the agility of the springbok; the stamina of the wildebeest; the rapier-like horns of the gemsbok – to quote but one example from each – all help individual animals to escape predation.

2.6 Relations between hyaenas and other carnivorous animals

Relations between carnivores are complex. Although a measure of overlap in diet exists amongst the large carnivores, direct competition for food is lessened through differences in prey selection and foraging habits. In addition, some of the smaller carnivorous animals also share common food interests with the larger ones, and some species, not necessarily the larger ones, may deprive others of food. Alternatively, one species may benefit from another without detriment to the benefactors. Finally, the predator hierarchy, where one species dominates another, even if food is not directly involved, may be important in inter-carnivore relations.

2.6.1 Brown hyaena and other carnivorous animals

2.6.1.1 LION
The remains of lions' kills are an important component of the brown hyaena's diet (Table 2.8), but are always scavenged after the lions have departed. If a brown hyaena locates lions on a kill, it will usually approach no closer than about 50 m down wind, look at the lions for a few minutes and then move off, returning intermittently until they have left the carcass. Sometimes brown hyaenas wait close by for the lions to finish. Then several may assemble at a carcass.

Lions may rarely compete with brown hyaenas for carcasses of animals that have died non-violently, and lions will also attack and can seriously maim brown hyaenas for no apparent reason (Eloff 1973b, Owens & Owens 1978, Apps 1982, section 4.1.5). In spite of these occasional attacks by lions on brown hyaenas, southern Kalahari brown hyaenas probably benefit more than they lose from the rather small lion population.

2.6.1.2 LEOPARD
Leopards have little influence on the brown hyaena population in the southern Kalahari and vice versa. The leopards' habit of taking their kills into trees, means that most of their food remains are unavailable to hyaenas (Fig. 2.23). On several occasions a brown hyaena came to a tree in which a leopard kill was hanging and spent several minutes sniffing around the base of the tree, sometimes picking up scraps that had fallen from the carcass.

On the few occasions that brown hyaenas and leopards met, they usually showed little interest in each other, merely keeping their distance. Occasionally, one may chase the other a short distance and I once found the remains of a six-month-old brown hyaena cub which had been eaten by a leopard. The following is a typical interaction between brown hyaenas and leopards:

9 September 1979. Cubitje Quap. 23.50 h. A male leopard is feeding on a scavenged wildebeest carcass when a brown hyaena approaches with its

Figure 2.23 A brown hyaena unsuccessfully attempts to reach the remains of a springbok stored in a tree by a leopard.

long hair and tail erect. The leopard retreats 20 m and lies down. The brown hyaena starts feeding and is soon joined by another one. The two hyaenas feed for an hour and a half while the leopard lies close by. When they leave, the leopard returns immediately. After a minute or two one of the hyaenas returns. The leopard continues feeding, growling a little, and the hyaena lies down 10 m away. The leopard drags the carcass away from the hyaena a few metres, at which the hyaena stands up and advances. The leopard drags the carcass even further away. The hyaena moves closer, the leopard stands, and the hyaena turns around and moves away.

Owens & Owens (1978) recorded a brown hyaena appropriating a kill from a leopard and on two occasions chasing a leopard into a tree.

2.6.1.3 CHEETAH
Circumstantial evidence suggests that cheetahs provided brown hyaenas with a significant amount of food along the Auob river bed. Four adult

female cheetahs with cubs, and at least two adult males, inhabited this region in 1979–1980, and were frequently seen with springbok kills. The carcasses were invariably eaten overnight by brown hyaenas.

Along the Nossob river bed the supply of cheetah-killed carcasses was more sporadic as the density of cheetahs was lower. Consequently, I have few direct observations of brown hyaenas scavenging from cheetah (Table 2.8). However, several times brown hyaenas scavenged springbok carcasses of unknown origin, and these may well have been cheetah kills.

Most of the scavenging by brown hyaenas on cheetah kills was done after the cheetahs had left the carcass voluntarily. Only one instance of a direct confrontration between these two species was seen:

28 July 1973. Langklaas. Just after dark a brown hyaena is seen circling down wind from a female cheetah and her three small cubs, which have just started to feed on a springbok carcass. The adult cheetah sees the hyaena and immediately runs towards it, while her cubs run off in the opposite direction. As the cat approaches the hyaena, it retreats a few metres, then turns around with mane and tail erect, and runs at the cheetah. They clash briefly, and the cheetah slaps and growls at the hyaena. Then the cheetah turns around and runs off in the direction of her cubs, abandoning the carcass to the hyaena.

Owens & Owens (1978) recorded similar interactions between brown hyaenas and cheetahs in the central Kalahari.

2.6.1.4 BLACK-BACKED JACKAL

Black-backed jackals are common in the southern Kalahari. They prey on springbok lambs, steenbok, springhares, and other small mammals, especially rodents, as well as taking birds, reptiles, arachnids, insects, wild fruits, and carrion (Bothma 1966, Smithers 1971, personal observations). Thus, their diet overlaps that of the brown hyaena.

On six occasions brown hyaenas chased jackals off kills; four steenbok, a springbok lamb, and a springhare; and three times a brown hyaena found a rodent buried under the sand, which had probably been cached by a jackal (Table 2.8). W. Ferguson (personal communication) twice saw a brown hyaena attempt to chase jackals off springhare kills, once with success, but the other time the jackals managed to escape with the remains. Brown hyaenas are alert to the activities of black-backed jackals, often moving over towards a jackal they sense, apparently to investigate the possibility of obtaining some food. They also readily eat jackals (Table 2.3).

Black-backed jackals are able to deprive brown hyaenas of a substantial amount of meat, by quickly finding a carcass and removing the remaining meat before the hyaenas arrive, although they are unable to eat the larger bones. Even when a hyaena is feeding on a carcass, the jackals are often able to continue feeding.

Black-backed jackals derive some benefit from brown hyaenas by picking up food scraps after a brown hyaena has eaten. Often, when a brown hyaena begins feeding on bones, one or more jackals will appear (possibly attracted by the sound of the bones being cracked), and wait until the hyaena moves away. Sometimes, by persistent pestering and even nipping the hyaena in the back legs, the jackals will cause the hyaena to move away a few metres with its food, and pick up the remaining scraps which the hyaena would probably have eaten. Jackals behave similarly towards spotted hyaenas (section 2.6.2.4). In these cases the presumably significant benefit derived by the jackal is hardly detrimental to the hyaenas.

Sometimes jackals will follow a brown hyaena for distances up to 2 km. The brown hyaena usually takes little notice, although if a jackal approaches too closely, it may chase it away. Jackals will also chase brown hyaenas (and other large carnivores) away from their dens during the breeding season, by following, barking, and nipping at the hyaena's back legs.

Relations between brown hyaenas and black-backed jackals in the southern Kalahari are again similar to those in the central Kalahari (Owens & Owens 1978). There is competition between the two species, and they both gain and lose food from each other. Because of the high density of black-backed jackals in the southern Kalahari, the most important impact of this relationship would seem to be the loss of food for brown hyaenas.

Figure 2.24 A feeding brown hyaena is closely attended by four black-backed jackals.

2.6.1.5 CARACAL

Caracals are comparatively rare in the southern Kalahari (Table 2.4), but brown hyaenas appropriated three steenbok, two springhares, and one each of a springbok adult, a springbok lamb, and an African wild cat (Table 2.8) – all containing a large proportion of meat – from caracals. Each time the hyaena merely moved quickly towards the caracal, which withdrew without putting up any resistance.

2.6.1.6 VULTURES

The white-backed vulture is common in the southern Kalahari. These birds are efficient scavengers, able to forage over an extensive area, and to rapidly dispose of the meat of a carcass (Houston 1974). They are, therefore, capable of depriving brown hyaenas of much food, particularly from carcasses of animals that have died non-violently. Vultures are diurnal and hyaenas are nocturnal. Apart from being a useful water conservation strategy, nocturnal behaviour of hyaenas lessens competition with vultures. This gives hyaenas a chance of finding animals that die late in the afternoon, or at night, before the vultures do.

As vultures are unable to chew bones, Mundy & Ledger (1976) and Richardson *et al.* (1986), have suggested that griffon vultures are dependent on hyaenas (mainly spotted) for bone fragments with which to supply their chicks with calcium. The vultures apparently return to carcasses after hyaenas have been feeding in order to obtain the bone fragments.

2.6.2   Spotted hyaena and other carnivorous animals

2.6.2.1 LION

Lions and spotted hyaenas are potentially serious competitors at food, as nearly 70% of their kills comprise the same two species: wildebeest and gemsbok. However, this competition is significantly lessened as they tend to select from different segments of these prey populations (section 2.4.4).

In East Africa the two species compete by robbing each other of their kills. In the Ngorongoro Crater lions took a significant amount of food away from spotted hyaenas (Kruuk 1972), but in a part of the Rwenzori National Park, Uganda, lions lost nearly a third of their carcasses to spotted hyaenas, although in most cases the lions had eaten most of the meat (Van Orsdol 1981). In the southern Kalahari spotted hyaenas obtain food by scavenging from lion kills (Table 2.13), mainly after the lions have departed, and vice versa (Fig. 2.25). Clashes over food between spotted hyaenas and lions are rare, with neither species having a clear advantage (Table 2.23), but they may be dramatic:

> 30 May 1980. 23.30 h. An adult female spotted hyaena is moving through the dunes east of Groot Brak, when she starts moving quickly upwind. After 2 km she stops and looks off at a lioness 100 m away lying next to a

Figure 2.25   Three young lions scavenge the remains of a spotted hyaena kill.

Table 2.23   Outcomes of interactions between spotted hyaenas and lions at food when one species in possession was challenged by the other.

Possessors (The opposite species from the challengers)	Challengers	
	Lions	Spotted hyaenas
Lose a substantial amount of food	1	2
Lose a little food	3	2
Maintain possession	2	3

less than half-eaten gemsbok carcass. The hyaena paces up and down, not approaching closer than 50 m from the lion. After 10 min she whoops 17 times, and before she has finished there is a reply. Within 2 min three other hyaenas from the clan arrive.

The hyaenas greet each other, while lowing (Table 5.6), and the lioness drags the carcass under a low-hanging shepherd's tree and starts eating. The hyaenas move slowly towards her, emitting a variety of lows, hoot-laughs, and whoops (Table 5.6), looking aggressive with their tails curled over their backs, the hair on their backs bristling, their heads up and their ears cocked (Table 5.1). As they approach her the lioness sits up to face them and growls, but the hyaenas continue to move forward. When they are about 5 m from her, the lioness lunges at them and growls louder, but this does not deter the hyaenas. A few seconds later she turns around and flees.

Even when no food is present spotted hyaenas and lions may clash with each other, as is illustrated in the following example:

5 July 1982. 23.35 h. A group of six spotted hyaenas come to Kousaunt windmill. They do not, as is normal, stop to drink, but lope past. They pass four gemsbok and ignore them, loping along at 15–20 km/h. After 2.7 km they suddenly change direction and, with their tails up, run hard after a lioness. The lioness flees and after 200 m jumps into a dead tree. The hyaenas gather at the foot of the tree looking up for a few seconds, then run off after a second lioness. This lioness stands her ground, growling, and the hyaenas surround her (Fig. 2.26a). She lunges at the hyaenas who jump back a short distance (Fig. 2.26b), before again advancing to within 5 m of her. After 2–3 min the hyaenas leave the lioness and go back to the first one who has come down from the tree. They surround her, but she moves towards them, and the hyaenas give way. The two lionesses come together, rub heads and face the hyaenas, which are now standing 5–10 m away. After a few seconds the lionesses back away some 10 m and lie down. Three of the hyaenas start whooping. After 3 min the lionesses move away and the hyaenas come up and sniff where they have been lying.

In interactions between spotted hyaenas and lions away from food (Table 2.24), spotted hyaenas were slightly more likely to initiate an interaction than the reverse. Interactions initiated by hyaenas, moreover, were much longer lasting than those initiated by lions, and when spotted

Table 2.24   Analyses of interactions between spotted hyaenas and lions away from food. A win was scored when one species caused the other to retreat.

	Spotted hyaenas	Lions	Uncertain
Initiators	9	7	4
Mean ± SE duration of interaction when initiated by:	57min ± 24	5min ± 2	-
Mean ± SE number of individuals in initiating group:	6.7 ± 0.8*	2.9 ± 0.9	-
Winners	5	10	5

* Mean spotted hyaena foraging group size was 3.0

(a)

(b)

Figure 2.26 Spotted hyaenas mobbing a lioness. (a) The lioness lies flat as the hyaenas approach, then (b) the hyaenas scatter as the lioness lunges at them.

hyaenas were the initiators, they tended to be in relatively large groups. Lions won most interactions, but spotted hyaenas won or drew all the interactions which they initiated. On three occasions (including the one recounted above) lions jumped into trees to escape the hyaenas.

Spotted hyaenas were far less likely to initiate an interaction (irrespective of the presence of food) when adult male lions were present (Table 2.25). The two instances when they did so were at carcasses. In one case the lions eventually moved away from the carcass when they had all finished feeding on it, and in the other the lions chased the hyaenas away. Spotted hyaenas were also unlikely to win when male lions were present (Table 2.25), and on two occasions hyaena cubs were killed by male lions (Fig. 2.27). The two occasions that spotted hyaenas won an interaction in which male lions were present were when a male failed to displace nine spotted hyaenas from two recently killed eland calves, and when the lions left a carcass prematurely. The presence or absence of male lions did not appear to influence the lions' inclination to initiate an interaction with the spotted hyaenas, nor its outcome (Table 2.25).

Neither exploitation nor interference competition for food between spotted hyaenas and lions appears to be important in regulating their populations in the southern Kalahari. Exploitation competition is lessened through differences in prey selection and interference competition is rare, because of the low densities of both predators. Most food scavenged from

Figure 2.27 An adult male lion shakes a seven-month-old spotted hyaena cub he has just killed.

Table 2.25　　Initiators and winners of all interactions between spotted hyaenas and lions if adult male lions were present or absent from the lion group.

		Adult males present	No adult males present	x^2
Initiators	Spotted hyaenas	2	14	7.56; d.f. = 1; $p < 0.01$
	Lions	7	5	0.00; d.f. = 1; $p > 0.05$
	Unknown	2	2	
Winners	Spotted hyaenas	2	8	2.50; d.f. = 1; $p > 0.05$
	Lions	7	9	0.06; d.f. = 1; $p > 0.05$
	No winner	2	4	

lions by spotted hyaenas is done so passively and consists of what the lions have left.

In spite of this, there is a large measure of aggression shown between the two species. This is particularly marked when spotted hyaenas initiate interactions with lions, and is not dependent on the immediate availability of food. This behaviour has been commented on by Kruuk (1972), and has been interpreted as mobbing behaviour, whereby the spotted hyaenas induce the lions to leave the area. It is arguable that it is to the spotted hyaenas' advantage to have the larger competitors out of the way.

Spotted hyaena–lion interactions in the southern Kalahari are not as one-sided in favour of lions as they are in the Serengeti or Ngorongoro Crater (Kruuk 1972, Schaller 1972). This is not due to relative differences in group sizes, as the ratio is more even in the Kalahari than in East Africa, where spotted hyaenas often occur in far larger groups than lions (Kruuk 1972). Neither does relative size appear to be important; although Kalahari spotted hyaenas are larger than their East African counterparts (Kruuk 1972, this study), the same is true for lions (Smuts et al. 1980). In an area where resources are often thinly and widely scattered, the loss of a kill to competitors may be more serious to the animals concerned than in an area where resources are more concentrated. Ngorongoro spotted hyaenas which lose a kill to lions may quickly be able to obtain another one, whereas Kalahari spotted hyaenas may have to travel 50 km or more before making the next kill. Being the smaller of the two species, spotted hyaenas may be more vulnerable to these losses than are lions. Therefore, it may be more worthwhile for spotted hyaenas in the southern Kalahari to take higher risks when attempting to steal kills from lions, when defending their own kills, or when attempting to get lions to move out of the area, than their counterparts in East Africa.

The contention of Packer (1986) that competition with spotted hyaenas is not an important advantage of group foraging in lions may not be valid. It can be argued that lions do not lose many carcasses to spotted hyaenas because they do forage in groups – solitary lions may not be able to compete with social spotted hyaenas.

2.6.2.2 LEOPARD

There is little exploitation or interference competition between spotted hyaenas and leopards in the southern Kalahari (Tables 2.13, 2.15, 2.22). On only three occasions did spotted hyaenas expropriate food from a leopard; a single hyaena chased a young female leopard from its springbok kill, four hyaenas took over a freshly killed ostrich, and four hyaenas took away a gemsbok calf. Bothma and le Riche (1984) recorded that a spotted hyaena drove a female leopard away from her kill, but that on two occasions a large adult male leopard kept two spotted hyaenas away from its kill. As mentioned earlier, the leopard's habit of taking its kills into trees precludes most scavengers from obtaining them. In the case where the hyaenas took away the gemsbok calf, the leopard had dragged the carcass some 150 m to the nearest tree after making the kill, but this was not suitable for storing a carcass in. It then ate the carcass underneath the tree until the hyaenas chased it away. After the hyaenas had left, the leopard placed the head, which was all that remained, in another tree 300 m away.

On nine occasions spotted hyaenas encountered single leopards away from food, and each time the hyaenas chased the leopard until it took refuge in a tree (eight times) or down a hole (once). On five of the occasions a single spotted hyaena was involved, on three, two hyaenas, and once there were five hyaenas.

Leopards, therefore, are usually dominated by spotted hyaenas, although the amount of food they lose is small.

2.6.2.3 CHEETAH

There is little overlap in prey selection between spotted hyaenas and cheetahs, and spotted hyaenas gain little food from cheetahs (Tables 2.13, 2.15, 2.21). Cheetahs are low in the predator hierarchy and it has been postulated that in the Serengeti spotted hyaneas deprive them of a substantial amount of food (Bertram 1979). I only observed spotted hyaenas drive cheetahs from their kill on five occasions, four of which were at night: once five hyaenas chased three cheetahs off a springbok, and several hours later chased them off a second one. A single spotted hyaena chased a cheetah from its freshly killed springbok lamb, six spotted hyaenas stole a steenbok carcass from a cheetah, and five hyaenas took the remains of a steenbok from a female cheetah and her four, nearly full grown cubs. In this last mentioned case the cheetahs retired 100 m from the hyaenas and lay on top of a dune. When the hyaenas had finished feeding they moved slowly, sniffing the ground, towards where the cheetahs were lying. As they approached them the cheetahs bolted to be pursued by the hyaenas, eventually taking refuge in two trees (Fig. 2.28).

Figure 2.28 Two of a group of five cheetahs take refuge in a tree from three spotted hyaenas.

On two occasions spotted hyaenas and cheetahs interacted in the absence of food: three hyaenas chased two cheetahs away from the hyaenas' den; a single hyaena moved over towards two cheetahs, which stood their ground and threatened, one lunging at the hyaena, which moved away.

The amount of food lost by cheetahs to spotted hyaenas in the southern Kalahari is small, but all five instances occurred within 5 km of hyaena dens. Cheetahs may on occasions be forced out of profitable foraging areas close to a spotted hyaena den, due to losses of kills to hyaenas.

2.6.2.4 BLACK-BACKED JACKAL

Overlap in prey selection between spotted hyaenas and black-backed jackals is small, and is chiefly confined to springbok lambs. Only once did a spotted hyaena displace black-backed jackals from a kill, when one ran

Figure 2.29 A black-backed jackal about to nip a spotted hyaena in the back leg while the hyaena feeds on the remains of a springbok.

300 m to where two jackals had just caught a springhare. Unlike brown hyaenas, spotted hyaenas do not often eat jackals.

As scavengers black-backed jackals may deprive spotted hyaenas of some food by being first onto a carcass, but this loss is more important to brown hyaenas (section 2.6.1.4). Black-backed jackals scavenge from spotted hyaena kills and show little fear of spotted hyaenas when feeding on a carcass, often darting in between them to steal a scrap. The hyaenas may chase them off a short distance, and nip and nudge jackals out of the way, but a hyaena was never seen to seriously hurt a jackal. As with brown hyaenas (section 2.6.1.4), when a single spotted hyaena is feeding, a jackal may harass it (Fig. 2.29), causing the hyaena to move, so that the jackal can obtain some scraps. Similar behaviour between jackals and spotted hyaenas has been reported by Kruuk (1972), and Bearder (1975).

2.6.2.5 VULTURES
This is not an important relationship for spotted hyaenas in the southern Kalahari, but as has been discussed in section 2.6.1.6, competition with vultures for scavenged food may be partially responsible for hyaenas being nocturnal, and bone fragments from hyaena-chewed bones may be vital to vultures. Vultures may deprive spotted hyaenas of some scavenged food and vice versa, and vultures may also be able to scavenge from spotted hyaena kills.

Table 2.26   A summary of the nature of the relationships between hyaenas and other carnivorous animals.

Species	On brown hyaena	By brown hyaena	On spotted hyaena	By spotted hyaena
Lion	+(-)	0	+(-)	0(-)
Leopard	0	0	0	-
Cheetah	+	0(-)	+	-
Black-backed jackal	-(+)	-	0	+
Caracal	+	-	0	0
Vultures	-	0(+)	0	+

+ Advantageous
0 Little or no effect
- Detrimental
() Occasional alternative nature of the relationship.

2.6.3   Conclusions

The relationship of the two hyaena species with the other carnivorous animals, in terms of the benefits or detrimental effects felt by the one on the other, are summarized in Table 2.26. Because it is almost exclusively a scavenger, the brown hyaena's relationships with the other carnivorous animals is different from the spotted hyaena's, in that brown hyaenas hardly provide any food for the others.

Only lions and cheetahs have an effect on spotted hyaenas, largely to the hyaenas' advantage, whereas all species, except leopards, affect brown hyaenas in one way or the other. Spotted hyaenas negatively affect all the large carnivores in some way, whereas brown hyaenas have little effect on the larger carnivores, but exert some negative effects on the smaller ones.

2.7   Summary

	Brown hyaena	Spotted hyaena
Diet	Opportunist Scavenged mammal remains, wild fruits – also reptiles, birds, eggs and insects	Specialist Mainly large and medium-sized mammals
How procured	Mainly scavenged (94% of biomass)	Mainly killed (70% of biomass), some scavenged
Prey selection	The occasional small animal	Mainly calves of large antelope, especially gemsbok, possibly less fit gemsbok adults
Consumption	0.07 kg of food/kg of hyaena/day	0.10 kg of food/kg of hyaena/day

Major competitors	Black-backed jackal	Lion
Major providers of scavenged food	Lion Non-violent mortality	Non-violent mortality Lion
Major beneficiaries	—	Vultures Black-backed jackal

Statistical tests

1. Comparison of the number of spotted hyaenas feeding from kills versus scavenged carcasses of the same species. Wilcoxon matched pairs test: $T = 80$, $N = 11$; $p < 0.05$; two-tailed.
2. Comparison of the observed versus the expected frequencies with which gemsbok calves were killed by spotted hyaenas. The expected frequency was calculated from the percentage calves counted in the population. $\chi^2 = 173.3$; d.f. $= 1$; $p < 0.001$; $N = 52$.
3. Comparison of the observed versus the expected frequencies with which wildebeest calves were killed by spotted hyaenas. The expected frequency was calculated from the percentage calves counted in the population. $\chi^2 = 22.8$; d.f. $= 1$; $p < 0.001$; $N = 16$.
4. Comparison of the observed versus the expected frequencies with which wildebeest sub-adults were killed by spotted hyaenas. The expected frequency was calculated from the percentage sub-adults counted in the population. $\chi^2 = 12.2$; d.f. $= 1$; $p < 0.001$; $N = 3$.
5. Comparison of the observed versus the expected frequencies with which wildebeest adults were killed by spotted hyaenas. The expected frequency was calculated from the percentage of adults counted in the population. $\chi^2 = 24.2$; d.f. $= 1$; $p < 0.01$; $N = 15$.
6. Comparison of the frequencies with which male and female adult wildebeest were killed by spotted hyaenas versus the frequencies with which they occurred in the population. $\chi^2 = 2.41$; d.f. $= 1$; $p > 0.05$; $N = 15$.
7. Comparison of the frequencies in which adult wildebeest in age-classes 4–6 and 7–9 (see Appendix C) were represented in the living population versus the frequencies with which they were killed by lions. $\chi^2 = 7.01$; d.f. $= 1$; $p < 0.01$; $N = 119$.
8. Comparison of the frequencies in which adult wildebeest in age-classes 4–6 and 7–9 were represented in the living population versus the frequencies with which they died non-violently. $\chi^2 = 3.04$; d.f. $= 1$; $p > 0.05$; $N = 127$.
9. Comparison of the frequency with which wildebeest were found killed by lions versus the frequency with which they died non-violently during the periods 1973–1977 and 1978–1982. $\chi^2 = 63.0$; d.f. $= 1$; $p < 0.001$; $N = 251$.
10. Comparison of the frequencies with which adult gemsbok in age-classes 4–6 and 7–9 were represented in the living population versus the frequencies with which they were killed by lions. $\chi^2 = 1.26$; d.f. $= 1$; $p > 0.05$; $N = 63$.
11. Comparison of the frequencies in which adult springbok in each age-class were represented in the living population versus the frequencies with which they were killed by cheetahs. $\chi^2 = 19.36$; d.f. $= 4$; $p < 0.001$; $N = 273$.

12. Comparison of the frequencies with which adult springbok in age-classes 5–6 and 7–9 were represented in the living population versus the frequencies with which they were killed by leopards.
$\chi^2 = 0.19$; d.f. = 1; $p > 0.05$; $N = 209$.

13. Comparison of the frequencies with which springbok adult males, adult females, sub-adults and lambs were killed by cheetahs ($N = 174$) versus leopards ($N = 43$).
$\chi^2 = 12.93$; d.f. = 3; $p < 0.01$; $N = 217$.

14. Comparison of the frequencies with which adult springbok in age-classes 5–6 and 7–9 were represented in the living population versus the frequency with which they died non-violently.
$\chi^2 = 6.72$; d.f. = 1; $p < 0.01$; $N = 220$.

Comparative foraging behaviour

The large differences in diet between the two species described in the previous chapter are reflected in comparably large differences in foraging and feeding behaviour. The brown hyaena is a solitary forager, investigating a wide variety of potential food sources and usually feeding on its own (Mills 1978a), whereas the spotted hyaena tends to concentrate on large and medium-sized mammals, most of which are consumed with conspecifics.

3.1   Activity patterns and day-time resting sites

Hyaenas are predominantly nocturnal (Fig. 3.1 and Table 3.1). This may be to avoid competition with vultures and to conserve water (section 2.6.1.6). An exception was three large brown hyaena cubs which were most active around sunset during the winter (Fig. 3.1).

Brown hyaenas were active for more of the 24h period than were spotted hyaenas (Table 3.1), particularly during the hours of darkness (18.00 h–06.00 h) (1). This is due to differences in foraging strategy between the two species, as brown hyaenas typically move continuously from one small food item to the next, whereas spotted hyaenas frequently feed on large food items and thereafter rest. On some nights spotted

Table 3.1   Time spent active by hyaenas from different areas.

Species	Area	Per cent of 24h period active	Per cent of hours of darkness active	Source
Brown hyaena	Kalahari	42.6	80.2	This study
Spotted hyaena	Kalahari	31.0	55.3	This study
Spotted hyaena	Kruger	27.5	-	Henschel 1986
Striped hyaena	Serengeti	26.0	53.0	Kruuk 1976a
Brown hyaena	Namib	15-37	-	Goss 1986
Spotted hyaena	Ngorongoro	16.0	-	Kruuk 1972

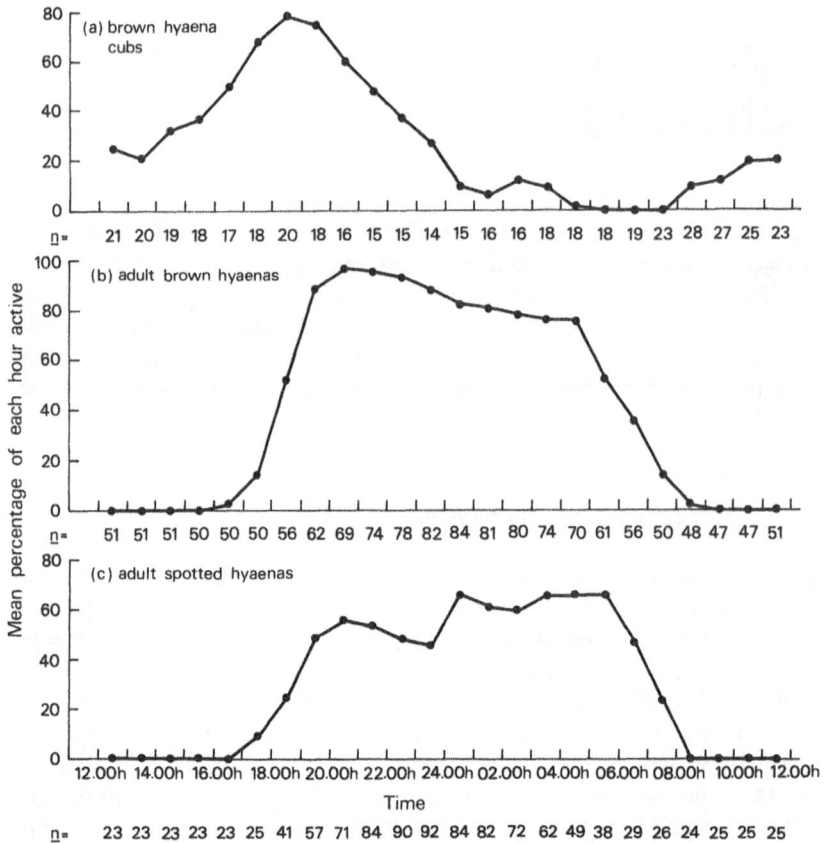

Figure 3.1 Mean percentage of each hour of the 24 h period that (a) brown hyaena cubs, (b) brown hyaena adults, and (c) spotted hyaena adults were active. Focal animal sampling was used, and the animal was regarded as being active if it was standing, walking or feeding.

hyaenas did not forage at all, whereas the least active a brown hyaena was, was for 62% of one night.

Ecological pressures also have an effect on activity budgets (Table 3.1). In the highly productive Ngorongoro Crater spotted hyaenas were less active than in other areas, and in the intermediately productive Kruger National Park they were active for longer than in Ngorongoro, but for less time than in the Kalahari. In a food-rich situation along the Namib Desert coast individual brown hyaenas were considerably less active than their Kalahari counterparts. In the Serengeti, the striped hyaena, which is ecologically similar to the brown hyaena, was found to be active at night for less time than were Kalahari brown hyaenas (2).

Figure 3.2 shows how southern Kalahari hyaenas allocated their time between four basic behaviour patterns. Spotted hyaenas spent relatively longer periods eating and socializing than brown hyaenas, and brown hyaenas spent more time foraging than spotted hyaenas.

The mean distance travelled per night by brown hyaenas (31.1 ± 2.1 km per night) was similar to that travelled by spotted hyaenas (27.1 ± 1.4 km per night) (Fig. 3.3) (3). The longest movement recorded by a brown

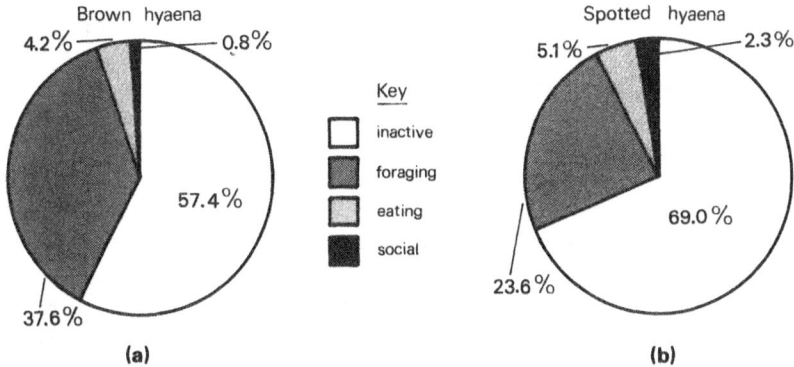

Figure 3.2 Relative time spent on various activities by brown hyaenas and spotted hyaenas. If a hyaena was moving it was regarded as foraging.

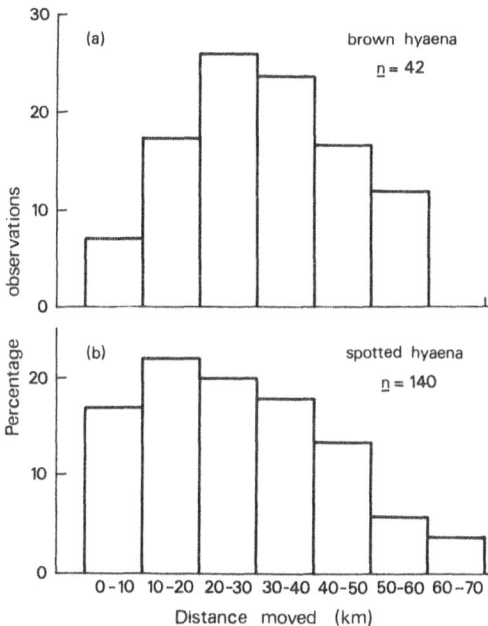

Figure 3.3 Distances which (a) brown hyaenas and (b) spotted hyaenas were observed to move when they were followed for a full night.

hyaena was 54.4 km, in 13 h 55 min (3.9 km/h), and the longest for spotted hyaenas was 69.1 km, undertaken by a group of five in 10 h 20 min (6.7 km/h). Brown hyaenas forage at a walk of about 4 km/h, they rarely lope, and then only for short distances. Their movements are erratic, punctuated by investigation of bushes, holes, trees, etc. Spotted hyaenas move less erratically, either walking at the same speed as brown hyaenas, or loping at a speed of 10 km/h. There was no difference in the distance that spotted hyaenas walked (53.3% of the distance sampled) as opposed to loped (46.7% of the distance) (4). Once, an adult male spotted hyaena covered 29.7 km in 3 h (9.9 km/h), and on another occasion two males covered 50.3 km in 5 h 30 min (8.6 km/h).

When resting during the day in summer, hyaenas must seek shelter from the heat. Brown hyaenas usually rest in a hole, or under a large shepherd's tree (Table 3.2). These trees are carefully selected, the hyaena often examining several before making a choice. An important criterion appears to be that the tree has thick branches close to the ground. A favoured summer daily resting site for spotted hyaenas is a drinking trough at a windmill. Alternatively, they often lie under a bush, and not always a particularly dense one. Before lying down both species dig a hollow in which to lie and during the course of the day they repeatedly kick sand with the back legs onto the stomach. As the temperature of the sand 25 mm below the surface is 10 °C lower than the temperature on the surface (Leistner 1967), this behaviour helps to cool them.

In the spring and autumn months the contrast in choice of daytime resting site is not marked between the two species and both usually choose a bush (Table 3.2). Although it is still hot in the day, the angle of the sun is more oblique, and a less dense tree or bush can provide adequate shade.

Table 3.2   Frequency of day-time resting sites used by brown hyaenas and spotted hyaenas at different seasons. The temperatures are the mean daily maxima for each season (Anon. 1986).

Resting site	Summer ± 35° C		Spring-autumn ± 30° C		Winter ± 23° C	
	Brown	Spotted	Brown	Spotted	Brown	Spotted
Hole	12(31%)	7(12%)	1(0%)	0(0%)	0(0%)	0(0%)
Large shepherd's tree	26(67%)	6(11%)	8(24%)	1(4%)	0(0%)	0(0%)
Large camelthorn tree	0(0%)	12(21%)	0(0%)	4(15%)	0(0%)	2(4%)
Water	0(0%)	15(26%)	0(0%)	0(0%)	0(0%)	0(0%)
Bush	1(3%)	17(30%)	24(73%)	22(81%)	27(61%)	25(56%)
Tall grass clump	0(0%)	0(0%)	0(0%)	0(0%)	17(39%)	8(18%)
Open	0(0%)	0(0%)	0(0%)	0(0%)	0(0%)	10(22%)
χ^2 (Brown vs. spotted)	53.64		13.06		0.64	
d.f.	4		2		1	
p	<0.01		<0.01		>0.05	

In winter, when keeping cool is no problem, the hyaenas utilize a different range of resting sites from those used at other times of the year (Table 3.2). However, at night they may be exposed to severe cold, with temperatures dropping to as low as $-10\,°C$ (Anon. 1986), and strong winds.

Skinner et al. (1984) have suggested that the long hair of the brown hyaena, in comparison with the spotted hyaena, provides more efficient thermal insulation during exposure to severe cold. In this respect the brown hyaena is better adapted to cold than is the spotted hyaena. Although spotted hyaenas are able to survive the cold spells in the southern Kalahari, they may only be able to achieve this by expending more energy than brown hyaenas. However, because of their longer hair, and in spite of the fact that they moult in summer, brown hyaenas may have to be more selective in choosing a cool resting site in summer (Table 3.2). Additionally, brown hyaenas may seek cover in order to escape detection while they are sleeping during the day. Unlike spotted hyaenas, brown hyaenas were never observed to rest under one of the large camelthorn trees along the river beds which, although good shade providers, leave the animals conspicuous.

3.2   Foraging group size

The extent to which animals move around solitarily or in groups, is an important aspect of their foraging behaviour and also has consequences for their social behaviour (Ch. 5). Over 99% of the observations of foraging brown hyaenas were of single animals (Fig. 3.4), and on the few occasions that two moved together they parted again after less than 2 km. Elsewhere too the brown hyaena is a solitary forager (Owens & Owens 1978, Goss 1986). In contrast, spotted hyaenas often forage in groups of varying size throughout their range (Fig. 3.5) (Table 3.3).

Differences in the food eaten by the two species seem to be important in determining foraging group size. Brown hyaenas scavenge most of their food and usually feed on small food items (section 2.2), so there is no advantage to foraging in a group. Southern Kalahari spotted hyaena foraging group sizes varied depending on which prey species they were hunting (Fig. 3.6). Springbok were hunted by significantly smaller groups (effectively by single hyaenas) than any other prey species, and gemsbok calves were hunted in smaller groups than gemsbok adults, wildebeest and eland, all of which were hunted in similar sized groups (Fig. 3.6 and Table 3.4). This is discussed further in section 3.6. Similarly, Kruuk (1972) showed that large groups of spotted hyaenas were most likely to hunt zebra, whereas smaller ones turned their attention to wildebeest or Thomson's gazelle. In the Timbavati Game Reserve and the Kruger National Park, impala are apparently hunted by spotted hyaenas solitarily

Figure 3.4 A brown hyaena forages along the Nossob river bed.

Figure 3.5 Five spotted hyaenas forage near Kousaunt.

Table 3.3   Foraging group sizes of spotted hyaenas from different areas. Data for Serengeti and
             Ngorongoro do not include single hyaenas.

Area	Mean	Percentage groups containing:				Source
		1	2	3 - 5	6 or more	
Kalahari	3.0	29.9	19.1	38.4	12.7	This study
Hluhluwe*	2.4	30.0	35.0	31.0	4.0	Whateley & Brooks 1978
Kruger	1.8	62.0	14.5	21.8	1.7	Mills 1985b
Namib	1.8	58.1	19.4	22.6	0.0	Tilson & Henschel 1986
Timbavati*	1.7	59.0	26.0	15.0	0.0	Bearder 1977
Ngorongoro*	-	-	22.0	39.0	39.0	Kruuk 1972
Serengeti*	-	-	46.0	39.0	15.0	Kruuk 1972

* Percentages approximations

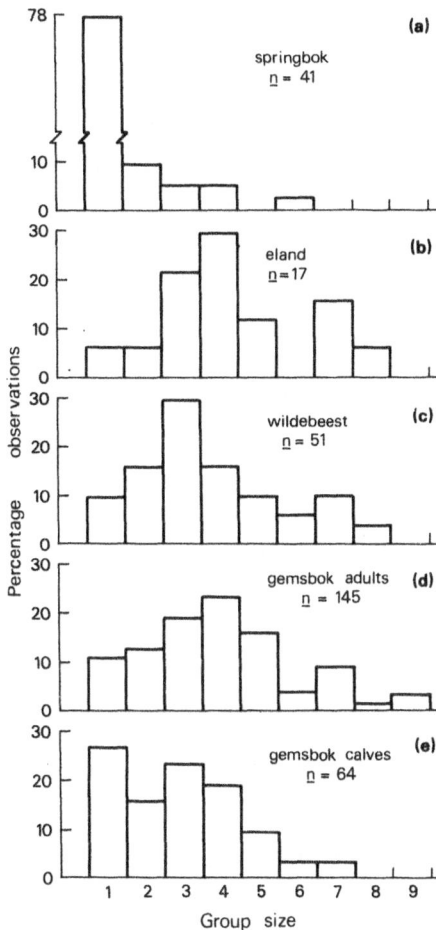

Figure 3.6 Group sizes of spotted hyaenas hunting various prey species.

Table 3.4   Two-tailed Mann-Whitney U tests comparing group sizes of spotted hyaenas hunting different prey species

	Gemsbok calf $n=64$	Gemsbok adult $n=145$	Springbok $n=41$	Wildebeest $n=51$	Eland $n=17$
Gemsbok calf	-	$U=3248.0$ $z=-3.5055$ $p<0.001$	$U=586.0$ $z=-5.0569$ $p<0.001$	$U=1233.0$ $z=-2.2866$ $p=0.02$	$U=312.5$ $z=-2.7355$ $p=0.006$
Gemsbok adult		-	$U=739.0$ $z=-7.46$ $p<0.001$	$U=3472.0$ $z=0.6565$ $p>0.05$	$U=1078.5$ $z=-0.8543$ $p>0.05$
Springbok			-	$U=263.5$ $z=-6.3877$ $p<0.001$	$U=60.0$ $z=-5.4679$ $p<0.001$
Wildebeest				-	$U=346.0$ $z=-1.2612$ $p>0.05$
Eland					-

or in twos, and larger species are mainly scavenged (Bearder 1977, Smuts 1979, Henschel 1986). This apparently precludes the need for spotted hyaenas to forage in large groups in these areas (Table 3.3).

3.3 The use of the senses during foraging

A description of the manner in which hyaenas use their senses while foraging is necessary for a complete description of foraging behaviour, and provides an example of how phylogenetic relations bring about similarities in behaviour which differs so markedly in other ways.

3.3.1 Smell

The most important sense used by hyaenas to find carrion is olfactory. The mean detection distance from downwind to six untouched carcasses by spotted hyaenas was 3.2 ± 0.4 km, the maximum 4.2 km. Even fairly old and dry carcasses can be detected by both species from up to 2 km downwind. However, if hyaenas are upwind of food, even a strong-smelling carcass, they are unable to detect it. Once, for example, a brown hyaena passed 250 m upwind of a fresh springbok carcass without noticing it.

Spotted hyaenas, like wolves (Mech 1970), are also able to detect live prey from downwind by smell, albeit from shorter distances than they are able to detect carrion (5). The mean detection distance by this method to 24 sources of potential prey was 1.1 ± 0.1 km, the maximum 2.8 km.

The intensity of the wind is also important in the detection of food by smell, and regulates the frequency and orientation of sniffs made by hyaenas. Brown hyaenas made significantly more sniffs when a steady wind was blowing than when no wind was blowing. Furthermore, with a steady wind all sniffs were upwind, whereas there were no differences in the orientation of their sniffs when no wind was blowing (Table 3.5). Then, however, hyaenas were able to detect a scent from all directions, although not from as far away as when favoured by the wind. They also have difficulty in pin-pointing the direction from which the scent originates. This is illustrated in Figure 3.7.

Apart from being able to detect scents through the air, hyaenas are also able to follow scent trails. This behaviour is characterized by the hyaenas moving with their noses to the ground, often circling around as they seemingly lose the scent, then picking it up again. I gained the impression that scent trails were easier to follow when the ground was damp. The mean distance scent trails were followed by spotted hyaenas until potential prey was encountered was 0.9 ± 0.2 km $(N = 11)$ and the maximum distance was 1.8 km, which is similar to the mean detection distance of live prey by scent through the air (6). Brown hyaenas too followed scent trails, the longest distance observed being 1.5 km to the remains of a springbok lamb. Furthermore, brown hyaena trapping success was enhanced if a carcass was dragged a few kilometres in different directions from a trap (section 1.3).

An impressive example of the spotted hyaena's ability to follow a scent

Table 3.5   Comparisons of (i) the frequencies with which brown hyaenas sniffed the air in a steady wind versus the frequency they did so with no wind, and (ii) the frequencies with which sniffs were made in one of four directions, or in several directions simultaneously with no wind blowing.

	Wind Conditions		χ^2
	Steady wind	No wind	
Distance followed (km)	95.4	30.2	
Total number of sniffs	292	63	
Sniffs /km	3.1	2.1	(i) 6.48; d.f.=1; $p < 0.05$
Number of upwind sniffs	292	-	
Number of other sniffs	0	63	
North	-	15	
South	-	18	(ii) 1.30; d.f.=4; $p > 0.05$
East	-	8	
West	-	14	
Several directions simultaneously	-	8	

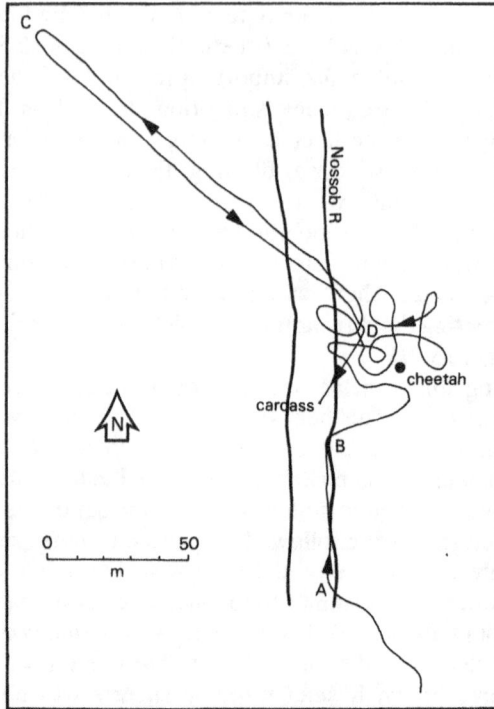

Figure 3.7 An example of a brown hyaena having difficulty in pin-pointing a carcass when no wind was blowing. At A the hyaena started sniffing, obviously having picked up a scent. At B it started circling around in a random fashion over an area of 0.5 km² where a cheetah was resting. After 45 min the brown hyaena abandoned the search and moved off in a northwesterly direction. After it had moved 1.4 km (C) a southwesterly wind started blowing and the hyaena immediately turned around and moved back towards the original area. When it reached the area it moved upwind quickly to a springbok carcass.

trail, as well as its ability to gain information from its conspecifics, is illustrated below:

6 September 1979. An old female of the Kousaunt clan arrives at the den at 24.30 h, having obviously come from a kill, as her face and fore-quarters are covered in blood and her belly is full. Another younger adult female and two sub-adults have been lying at the den for several hours. Immediately, the younger female moves off in the direction from which the older one had come, sniffing along the ground. However, she returns to the den after going some 500 m and lies down again with the others. After 10 min she stands up again, moves over to the older one, sniffs her on the face, and again moves off in the direction from which the older animal had appeared. This time the two sub-adults follow her.

The three hyaenas continue across an open plain sniffing the ground from time to time, moving cross-wind of the light breeze that is blowing.

After they have moved 6 km they start sniffing the ground, circling quickly, then moving forward again, still cross-wind. They meet up briefly with another hyaena which has also obviously eaten. After moving for another 2 km they stop again, sniff the ground, and after running back and forth a short distance, come to a wildebeest carcass.

3.3.2 Hearing

Hyaenas have an acute sense of hearing which often enables them to find food they could not have smelt.

On nine occasions a spotted hyaena which had been lying down for some time, suddenly stood up, cocked its ears in a certain direction for a few minutes, then ran off in that direction to join others feeding on a carcass. I presume it had heard them vocalizing. The mean distance of detection by this manner was 4.2 ± 1.2 km, with the two longest distances observed being 10.5 km and 10.0 km.

Hearing was also used in other situations. On nine occasions spotted hyaenas heard live prey (either gemsbok clashing horns, or large wildebeest herds vocalizing) from a mean distance of 2.4 ± 0.7 km and ran over to investigate. One wildebeest herd was detected from 6.0 km away. Other carnivores were also occasionally heard feeding by hyaenas, which then ran over to steal the food. A brown hyaena crunching the bones of a springbok carcass was detected by a spotted hyaena from 1.1 km, and two jackals which were feeding on a springhare were heard from 300 m by a brown hyaena.

The cry of a springhare under attack is a particularly strong stimulus for brown hyaenas. I sometimes used this to attract brown hyaenas by playing a tape-recording of the sound. The 'mobbing' call of the black-backed jackal, which is usually directed at other carnivores, is another. For example, a brown hyaena ran 700 m to where two jackals were calling at a leopard, which had a springbok carcass in a tree.

On occasion, both species employ a lie-and-wait foraging strategy, using sound as the cue. This was seen when large concentrations of springbok with lambs were in an area. Usually a single hyaena would lie down close to the springbok, every so often lifting its head and cocking its ears, as if listening for any disturbance. Once, spotted hyaenas scavenged a springbok from cheetah in this manner. Similar behaviour was observed in East African spotted hyaenas by Kruuk (1972).

3.3.3 Sight

Sight is of particular importance during hunting. Hyaenas obviously have good night vision; at night they would often stop and look at things which I could not see without the aid of a light. Brown hyaenas usually stop and/or look at prey before chasing, and spotted hyaenas were judged to have visually detected potential prey in 76 (60.3%) of 126 contacts in flat, open river habitat. Sight is also the means by which a potential victim is selected and followed during the chase.

3.4   Foraging for vegetable matter, birds' eggs and insects

Brown hyaenas utilize fruits, fungi, birds' eggs and insects to a greater degree than do spotted hyaenas (Ch. 2). Consequently, they differ in the amount of time and energy they spend foraging for these foods, and in the ecological impact each species has on the food type.

3.4.1   Vegetable matter

The two fruits eaten most often by brown hyaenas, the tsama and gemsbok cucumber (Table 2.5), grow in scattered patches of varying size (Fig. 2.1), their distribution and abundance depending, at least partially, on rainfall. In years of low availability brown hyaenas apparently learn where the major fruit patches are in their territory, and make special trips to these. Similar behaviour has been described by Kruuk (1976) for striped hyaenas and their favourite fruit tree, the desert date. The mean number of tsamas eaten during each feeding bout was 2.0 ± 0.18 and the mean number of cucumbers was 1.4 ± 0.11, although on most occasions only one fruit was eaten during each bout (Fig. 3.8). The fruits are bitten open with the canines (Fig. 2.5), and the flesh is scraped out using the incisors and tongue, so that when the hyaena has finished eating only a few fragments remain.

Brown hyaenas may be important agents in the dispersal of tsama and gemsbok cucumber seeds, as those that are ingested pass out undamaged in their faeces. The seeds are sought-after food items for rodents, and may be protected from rodents in the hyaenas' faeces. Tsama patches often grow around old brown hyaena dens and at latrines.

Spotted hyaenas eat tsamas more superficially than do brown hyaenas, ingesting a few small pieces of flesh before moving on. Spotted hyaenas, therefore, have little influence on the dispersal of the seeds of these fruits, which in turn have little influence on their movements.

Brown hyaenas also spend time feeding on smaller fruits and other vegetable matter: eight feeding bouts lasting from 15 s to 15 min were observed on brandy bush berries, the hyaenas plucking the small berries with their incisors. On three occasions a brown hyaena was observed to feed on the small fruits of the prostrate annual merremia for 10, 15 and 34 min. This plant grows in large patches, and the hyaena moved slowly and erratically through a patch with its head down, picking off individual fruits.

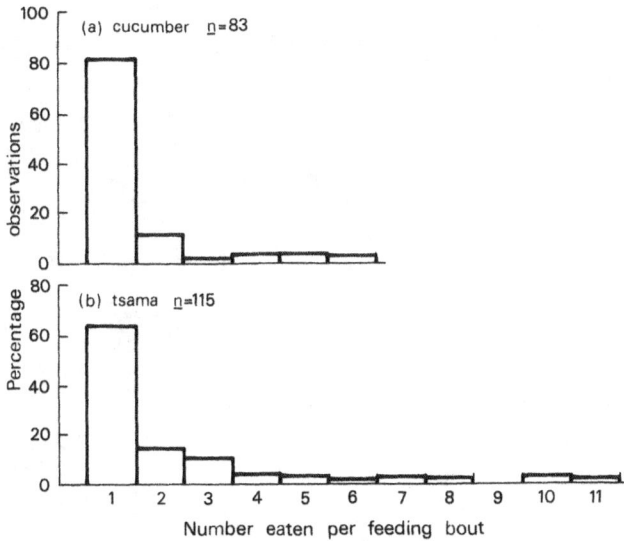

Figure 3.8 Number of (a) gemsbok cucumbers and (b) tsamas eaten by brown hyaenas during each feeding bout.

Finally, in April 1975 a brown hyaena dug out and ate 21 Kalahari truffles in two nights. On at least seven occasions the digging was preceded by an upwind turn of up to 200 m. This fungus, related to the European truffle, occurs in calcareous, sandy soils, a few centimetres below the ground (Leistner 1967).

3.4.2 Birds' eggs

Ostrich eggs are prized food items for both species, but particularly for brown hyaenas. Occasionally, a hyaena will encounter a single abandoned ostrich egg and eat it. Once, as a brown hyaena bit open such an egg there was a loud explosion, but the unpleasant odour emitted did not deter the hyaena from eating the entire contents.

More significant are the observations of both species raiding ostrich nests, although once again there are differences in how they accomplish this. The only time a brown hyaena was observed to raid an ostrich nest, it cautiously followed a scent trail for half an hour, before finding an unattended nest with 26 eggs (Fig. 3.9). (The behaviour of the brown hyaena at the nest is described in section 3.7.2.1.) In contrast, spotted hyaenas confronted ostriches sitting on nests. Once, four hyaenas killed a cock at the nest and also ate some of the eggs. On another three occasions, spotted hyaenas chased ostriches off their nests for distances varying from 100 m to 3 km, returning to the nests to eat the eggs.

Figure 3.9 A brown hyaena stands at a newly discovered ostrich nest. Note the ear notch for individual recognition.

Brown hyaenas have no difficulty in eating ostrich eggs. They bite a small opening in the top using the premolars, then lap up the contents with the tongue (Fig. 3.10). As the level in the egg drops they bite the egg open further, continuing in this way until they have finished, leaving the empty shell. Any of the contents that spill onto the sand are lapped up.

Spotted hyaenas, on the other hand, have difficulty in biting open ostrich eggs. At a nest the hyaenas will attempt to kick the eggs backwards with their forefeet, although most 'kicks' miss the egg and merely kick up a shower of sand (Fig. 3.11). Sometimes they connect, and eventually one egg will collide against another and crack. Immediately it hears the cracking noise, the hyaena goes to the egg and bites it open. On one occasion a spotted hyaena found a single egg and also repeatedly tried to kick it backwards, although there was no other egg or hard object close by on which to crack it. Eventually, 30 min after finding the egg, it managed to bite it open.

Brown hyaena cubs also kick ostrich eggs backwards, but without success. Similar behaviour with eggs has been described in several carnivores, for example in golden jackals (Kruuk 1975b), some mustelids (Van Gelder 1953, Rowe-Rowe 1978a), and particularly in viverrids (Ewer 1973, Eisner & Davis 1967), where eggs and other hard food items are thrown through the back legs against a hard object such as a stone, and broken.

Why brown hyaenas should be able to bite open ostrich eggs with relative ease, and even carry them in their mouths (one carried an ostrich

Figure 3.10 A brown hyaena eats an ostrich egg.

Figure 3.11 Spotted hyaenas at an ostrich nest. The animal on the right is attempting to kick an egg backwards with its forefoot. The one on the left is opening an egg after managing to crack it by kicking it against another one.

egg for 6.8 km), while the larger spotted hyaenas and even lions (D. Hughes personal communication) find difficulty in accomplishing these things, is not clear. None of the more obvious hypotheses provide the answer; for example, spotted hyaenas at any rate have as powerful jaws as brown hyaenas, and the shape and number of teeth in the two hyaena species are similar.

3.4.3   Insects

Spotted hyaenas eat few insects (Table 2.11), and although they form an important component of the brown hyaena's diet (Mills & Mills 1978) (Table 2.7), I have few direct observations of brown hyaenas eating them. As suggested in section 2.2.1, many of the unidentifiable food items eaten by brown hyaenas were probably insects. Positive identifications include: a locust, snapped off a tall grass stalk as the hyaena passed it; some dung beetles picked off the ground; a brown hyaena moving slowly in an erratic manner for several minutes with its nose to the ground, picking up some large black ants; and three cubs snapping erupting termites out of the air and off the ground. Brown hyaenas were never observed to run and then to jump into the air to catch a flying insect as were striped hyaenas (Kruuk 1976).

The only insects which were easy to observe being eaten were harvester termites. Twelve feeding bouts by brown hyaenas and one by a spotted hyaena were observed on these insects. Two lasted for less than 0.5 min, seven (including the spotted hyaena's) from 1–5 min, two for 10 min, one for 26 min, and one for 48 min. Termites were located fortuitously; for example, a brown hyaena cub was lying on the mound in front of its den when it stood up with its ears cocked, and moved quickly upwind to a bush around which was a colony of termites. When feeding a hyaena walked slowly amongst the termites with its nose close to the ground and its ears cocked, licking them up. I could not determine what caused the termination of a feeding bout. It was not caused by a chemical mode of defence, as suggested by Kruuk & Sands (1972) for aardwolves feeding on snouted harvester termites, as harvester termites do not secrete distasteful chemicals. Richardson (1985), suggested that aardwolves stop feeding on harvester termites when the density in the patch falls below a certain threshold, i.e. the marginal value (see Krebs 1978). This may also be true for hyaenas.

3.5   Brown hyaena hunting behaviour

The largest difference in foraging behaviour between the two hyaena species is their hunting behaviour. Hunting occupies a small part of the brown hyaena's foraging budget, whereas spotted hyaenas spend much time and energy in the pursuit of prey. But before analyzing hunting behaviour, it is important to establish what a hunting attempt is.

Several authors, including Mech (1970), Schaller (1972), and Kruuk (1972, 1976), have emphasized that considerable caution must be exercised when analysing hunting behaviour, as it depends so much on what is defined as an attempt. For example, hyaenas sometimes run at potential prey, making them run, possibly to assess if an individual is unfit and thus a suitable quarry. It is sometimes difficult to judge if the hyaena is testing in this manner, or if it is chasing with the intent to kill. Furthermore, brown hyaenas, in particular, sometimes chance upon a small animal which they try to catch by lunging at it, or by giving chase. Here again it is often difficult to decide if the act should be included as a hunting attempt. Nor does it help to include only those cases where the hyaena ran more than a certain minimum distance after a prey species, as the distances that brown hyaenas chased prey in successful hunting attempts varied from 1–1100 m (Table 3.6). I have taken a broad view and defined a brown hyaena hunt as any interaction between a brown hyaena and potential prey, where the hyaena moved towards the prey at an increased speed, provided that there was no carrion in the vicinity.

3.5.1 Mammals

Brown hyaenas rarely hunted small mammals, and only six (4.7%) of 128 attempts were successful (Table 3.6). (The large number of unknowns in

Table 3.6   Results of aspects of hunting attempts on mammals by brown hyaenas.

Mammal	Number of attempts	Number successful	Chase distances(m)		
			Mean	Standard error	Range
Bat-eared fox	16	1	111	18.9	5 - 200
Springbok adult	5	0	130	33.9	50 - 150
Springbok lamb	17	1	162	39.4	1 - 1100
Springhare	18	2	87	20.5	10 - 350
Black-backed jackal	3	0	23	13.3	10 - 50
Steenbok	3	0	303	170.3	10 - 600
Small rodent	2	1	10	-	3 - 20
Duiker	1	0	50	-	-
Aardwolf	1*	0	-	-	-
African wild cat	1	0	150	-	-
Honey badger	1	0	15	-	-
Hare	1	0	45	-	-
Striped polecat	1	1	10	-	3 - 20
Unknown	58	0	92	13.6	1 - 500
Total	128	6	-	-	-

* Attempted to dig the prey out of a hole

the table is due to the fact that, on dark nights and in tall grass areas, I was often unable to discern what species of prey the hyaena was chasing.)

Large herbivores were never hunted by a brown hyaena. The flight distance of large herbivores from a brown hyaena is short and it appeared that brown hyaenas were usually more concerned in avoiding them than the reverse, although generally they ignored each other. Occasionally, if a large herbivore or group of herbivores was encountered by a brown hyaena, it would stand for some time looking at them, and then would move on, often changing course slightly so as to avoid them.

Some details of brown hyaenas hunting specific mammals are given below:

3.5.1.1 SPRINGHARE

Springhares were abundant nocturnal mammals in the study area (section 2.1.2), and were the most common species hunted by brown hyaenas (Table 3.6). Of 18 hunting attempts on springhares, only two were successful. In addition, many of the unidentifiable prey animals chased were probably springhares. Usually the hyaena gave up the chase once it became obvious that the springhare was too far ahead to be caught, but on three occasions the springhare escaped into a hole, and on one of these the hyaena dug at the hole for nearly 5 min before giving up. One of the springhares (Fig. 3.12) was caught after a chase of not more than 20 m, and the other after the only hunting attempt observed involving two brown

Figure 3.12   A brown hyaena with the remains of its springhare kill.

hyaenas. After being chased for 150 m the springhare doubled back past the front hyaena and was easily caught by the second, 20 m behind. This animal ran off with the springhare, pursued by the other, which gave up after 300 m, leaving the hyaena which caught the springhare to eat it alone.

3.5.1.2 SPRINGBOK

Brown hyaenas and springbok herds generally ignored each other. Occasionally, a brown hyaena ran slowly towards a herd of springbok, causing them to run off a short distance, but on only five occasions did one pursue adult springbok, for distances varying from 50 to 150 m, without ever getting close to them. Once, a brown hyaena foraging in steady rain moved rapidly upwind to where an adult springbok was lying. As the springbok ran off, the hyaena gave a determined chase for 50 m, after which it appeared to lose sight of the springbok. During the springbok lambing season brown hyaenas showed more interest in springbok as is illustrated below:

> 7 January 1976. 20.05 h. A female brown hyaena is moving erratically, often with her nose to the ground, amongst a large herd of springbok on the limestone plains near Kwang. A lamb gets up 2 m away from her and runs off. The hyaena lunges at it, but does not pursue it. A little later she gives chase to another lamb for 50 m, but the lamb easily escapes. She practically stumbles on a springbok lamb which runs off, and again the hyaena lunges at it, but does not pursue it. Then she runs off quickly after a springbok lamb over a rise and, when I find her again 300 m further on, she is standing looking around. Further on she turns upwind, moving at a fast walk, then runs for 15 m as a lamb "pronks" (see section 3.6.4) away, and sniffs around where the lamb had been lying. She stays among the springbok until 23.00 h, and chases two more lambs, one for 60 m, the other for 20 m, before moving on.

Except for the two cases when the hyaena merely lunged at a lamb, these interactions were regarded as five hunting attempts. On one occasion a brown hyaena approached a lying springbok lamb from downwind. As it drew nearer it moved slowly with its head held low, until it was not more than 2 m away from the lamb. At this point the lamb ran off, chased hard by the hyaena for 150 m, before the hyaena gave up. Similar stalking behaviour in the brown hyaena has been recorded by Goss (1986), and in the striped hyaena by Kruuk (1976). The only successful hunt of a springbok lamb occurred when a brown hyaena flushed a lamb and chased it for 1100 m at a speed of approximately 30 km/h.

3.5.1.3 BAT-EARED FOX

On 15 occasions a brown hyaena was judged to have made an attempt to catch a bat-eared fox by chasing it, and on one occasion it was successful (Table 3.6). Twice, the fox ran into a hole, and on the other occasions the brown hyaena gave up after chasing for distances ranging from 5–200 m, often in a series of large circles. The sixteenth hunting attempt occurred when a brown hyaena dug for 7 min at a bat-eared fox den, and then gave up. The hunt was initiated when a hyaena encountered foxes, either after turning upwind a few metres, or after it had stopped and looked towards the foxes with its ears cocked. During the chase any other foxes close by were attracted to the activity, sometimes running over to the hyaena and trying to distract it by following it with the tail held vertically in the agonistic U posture described by Kleiman (1967). When a hyaena abandoned the hunt the foxes followed it closely, jinking around it, sometimes uttering a high-pitched bark, and causing it to run off with its hair raised. On nine other occasions bat-eared foxes followed a brown hyaena in a similar manner, even though the hyaena had made no attempt to chase them. Similar behaviour by bat-eared foxes towards striped hyaenas has been described by Kruuk (1976), and I once observed four bat-eared foxes follow a leopard in this manner until it jumped into a tree. These are further examples of mobbing behaviour described in section 2.6, having the effect of causing the hyaena to leave the area.

3.5.1.4 STRIPED POLECAT
Three interactions between brown hyaenas and striped polecats were observed. A large cub ran 10 m to catch a striped polecat, biting it in the head to kill it (Fig. 3.13). The polecat was at the entrance to a small hole into which it presumably had tried to escape. An adult hyaena sniffed at a

Figure 3.13 A large brown hyaena cub carries a striped polecat it has just killed.

bush out of which jumped a polecat, which the hyaena then left alone. A third hyaena lunged half-heartedly at a polecat which turned to face the hyaena, repeatedly approaching and backing away from it with its tail held up and curved forward over its back in the threat display (Rowe-Rowe 1978b), before turning around and running away.

In these latter two incidents the hyaena was in a good position to catch the polecat, yet made no attempt to do so. The cub that caught the polecat was one of three whose mother was dead and was possibly food-stressed. On the other hand, the adults in the above examples were enjoying a good supply of food when these incidents took place. As only 0.3% of the brown hyaena scats contained striped polecat hair (Mills & Mills 1978), it seems that striped polecats are not highly sought after food items, and may be eaten only when other food is scarce. Alternatively, anti-predatory behaviour such as chemical defence (see section 3.6.7.5) might be highly efficient in polecats warding off predators, although in neither case did I detect a smell.

3.5.1.5 OTHER MAMMALS

On three occasions a brown hyaena unsuccessfully chased a black-backed jackal for a short distance. Three unsuccessful hunting attempts on steenbok were observed; one a brief, hard chase of only 10 m, the other two more determined chases of 300 m and 600 m. Once, a brown hyaena caught a small rodent after a short chase in circles, and once, one escaped under a bush. A brown hyaena chased an African wild cat for 150 m before it took refuge in a tree. A honey badger escaped under a bush after a short chase of 10–20 m, from where it faced the hyaena and emitted a grating, rattling sound as described by Sikes (1964). On four occasions a foraging brown hyaena encountered porcupines, but the hyaenas showed little interest in the porcupines, and the porcupines in turn were not unduly concerned by the hyaena.

From spoor it was possible to reconstruct chases after a hare and a duiker, and once a brown hyaena was tracked to an aardwolf den where it tried to dig out two large cubs. On four other occasions based on direct observations, three after hard chases of 20, 100 and 150 m, the other when the hyaena suddenly sprang to its right, a hyaena dug briefly at fairly large holes. These too might also have been aardwolf dens.

3.5.2 Birds

Twenty-three observations were made of brown hyaenas attempting to catch birds (Table 3.7). Apart from kori bustards and adult korhaans, these were chance encounters, when a hyaena flushed a bird and lunged at it. On two occasions a brown hyaena snapped up a small bird, and on another it flushed a black korhaan with two chicks and ate them.

Once, a brown hyaena ran 25 m towards a kori bustard, approximately

Table 3.7   Results of aspects of hunting attempts made on birds by brown hyaenas.

Bird	Number of attempts	Number successful	Chase distances (m)		
			Mean	Standard error	Range
Korhaan	9	4	6	1.2	1 - 12
Black korhaan chicks	1	1	-	-	-
Unidentified small bird	5	2	-	-	-
Kori bustard	3	0	12	7.0	1 - 25
Cape dikkop	3	0	-	-	-
Crowned plover	1	0	-	-	-
Sandgrouse	1	0	-	-	-
Total	23	7	-	-	-

60 m away, which flew off, and on two other occasions short chases were also made, but again the bustards flew off. On nine occasions a hyaena stopped, turned sharply (I was unable to determine if this was upwind), and then moved quickly with its head down towards a korhaan (black or red-crested), which either flew away (five occasions), or was caught (four occasions).

Previously (Mills 1978a), I suggested that a certain brown hyaena had developed a technique for catching korhaans, as until then it had been the only one to hunt these birds. Subsequently, I have observed two other brown hyaenas hunting korhaans in the same way. This trait, therefore, appears to be more widespread in the southern Kalahari brown hyaena population than previously thought.

3.5.3 Some evolutionary considerations of the brown hyaena's hunting behaviour

The hunting behaviour of the brown hyaena in the southern Kalahari is similar to that described by Owens & Owens (1978) for the brown hyaena in the central Kalahari and by Kruuk (1976: 103) for the striped hyaena; i.e. 'a primitive chase and grab affair', where mammalian and bird prey form a small contribution to the diets of both species. This contrasts markedly with the specialized hunting behaviour of the spotted hyaena (Kruuk 1972, section 3.6).

The only time brown hyaenas purposefully looked for potential prey, was when hunting springbok lambs. Springbok lambs constitute the largest overlap in prey selection between brown hyaenas and spotted hyaenas in the southern Kalahari (section 3.6), but even here the overlap in prey interests is small. Brown hyaenas rarely hunt springbok lambs, and their success rate when they do is only 6%, compared with a 31% success rate in the case of spotted hyaenas (section 3.6.4).

Although I only once observed two brown hyaenas hunting together, I did get the impression that this enhanced success. However, only one of the hyaenas derived any benefit from the hunt, and it is clearly not beneficial for brown hyaenas to hunt small animals in groups, particularly as they hunt so rarely. Moreover, this hunt was a chance occurrence as the two hyaenas were indulging in social activities, and were certainly not foraging seriously. Only when one of them saw the springhare approximately 20 m away did the chase start.

3.6 Spotted hyaena hunting behaviour

In this section I present my observations on the spotted hyaena's hunting behaviour, and the anti-predatory behaviour of their prey. An important aspect here is the relationship between hunting success and hunting group size.

3.6.1 Gemsbok calves (age-classes 1 and 2)

25 August 1980. 02.00 h. Three spotted hyaenas are walking through the dunes near Groot Brak. Suddenly they start running upwind, their noses in the air. They stop, sniff the ground, then change direction slightly, running off over a dune with their tails curled over their backs. In the next dune valley, they come running up to a herd of seven gemsbok scattered over an area of about 30 × 20 m. At least one of the gemsbok is a calf of approximately ten months old, and the hyaenas immediately run after it, one hyaena slightly behind the other. The calf rapidly gains distance from the hyaenas which are running at approximately 50 km/h. After 1.3 km the calf starts turning to the left and the hyaenas gain on it slightly by cutting the corner. The calf continues to circle and the hyaenas begin to gain more as they again cut corners. After 1.7 km the hyaenas catch up with the calf. They pull it onto its side, biting at the stomach. The calf bleats and tries to lift its head, and eventually manages to roll onto its brisket. The hyaenas continue pulling out its innards (Fig. 3.14), and within two minutes it is dead.

The above is a typical example of a gemsbok calf hunt. The hyaenas located the prey by scent, quickly selected a calf, and then pulled down their victim after a long and fast chase. Towards the end of the chase, presumably as the calf tired, it began to circle, and this was where the hyaenas gained the upper hand.

It was difficult to see exactly what happened when hyaenas first encountered a herd of gemsbok, the poor light conditions and speed with which things happened made detailed observations impossible. It seems that often a gemsbok herd is dispersed at night, with the calves frequently

Figure 3.14 Two spotted hyaenas, with a third looking on, pull out the innards of a ten-month-old gemsbok calf after chasing it for 1.7 km.

lying down, while the adults feed close by. As soon as spotted hyaenas are detected, the gemsbok attempt to bunch together so that the calves can be defended. If the gemsbok do manage to close ranks around the calves, the hyaenas are unlikely to be able to cut a calf out of the herd (this occurred twice) due to the determined efforts of the adults, which repeatedly challenge the hyaenas with their rapier-like horns. Defence by adults was successful in 25% of the interactions between spotted hyaenas and gemsbok with calves (Fig. 3.15).

The outcome of the 63 observed interactions between spotted hyaenas and gemsbok calves are schematically summarized in Figure 3.15. Hyaenas abandoned the chase in only nine (18%) of the cases; three times after distances of under 1 km, and in the rest after distances varying between 2 and 4.4 km. The instances where the hyaenas abandoned the chase all involved large gemsbok calves, where it seemed that the hyaenas were outrun by the gemsbok. In 40 (82%) of the chases the hyaenas managed to catch and kill a calf. This gives an overall success rate of 63% of all encounters between spotted hyaenas and gemsbok calves.

Figure 3.16 illustrates four successful gemsbok calf hunts, showing some of the ways in which calves were separated from the adults, and how the calf would typically start to circle.

The average distances spotted hyaenas had to run before catching gemsbok increased with the age of the gemsbok up to sub-adulthood (7),

55
Hyaenas encounter gemsbok calves with adults

8
Hyaenas fail to cut out a calf
from the herd

6
Hyaenas cut a calf out but it
gets back to adults

Adult
defence

41

8
Hyaenas encounter calves alone

49
Calves are chased

3
Hyaenas give up within 1000m

46
Calves are chased > 1000m

6
Hyaenas give up after a
mean distance of 3.0km

40
Calf killed after a mean distance
of 1.7 km

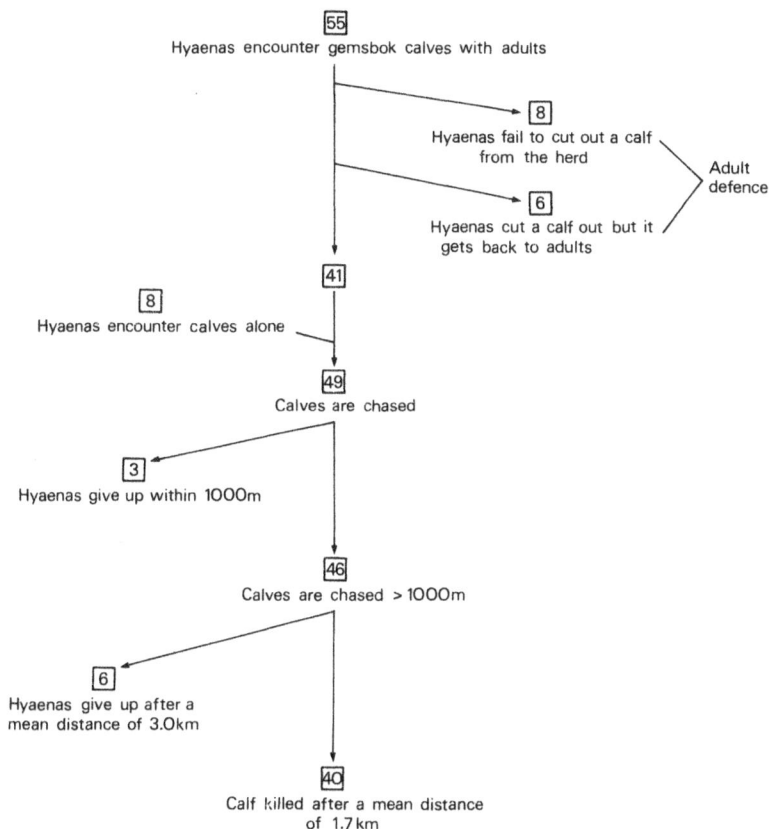

Figure 3.15 Results of interactions between spotted hyaenas and gemsbok herds in which there were calves. Boxed figures indicate those interactions considered as hunting attempts.

after which the average dropped (Fig. 3.17). Excluding new-born calves, the average distance gemsbok of age-classes 1–2 were chased before being caught was 1.6 ± 0.2 km (range 0.25–3.8 km). It was impossible to measure accurately the speed of these chases, but they usually reached 40–50 km/h.

One of the crucial factors in a gemsbok calf hunt is whether the hyaenas can get a calf away from the adults. To this end the number of hyaenas hunting, and the number of gemsbok in the herd, might be important factors. However, Figure 3.18 shows no relationship between hunting success on gemsbok calves and the number of gemsbok in the herd, and Figure 3.19 shows no significant relationship between hunting success and the number of spotted hyaenas in the hunting group. Nor was there any indication that one hyaena hunting on its own was any less successful than two or more hyaenas hunting together (8). Even groups of five or more

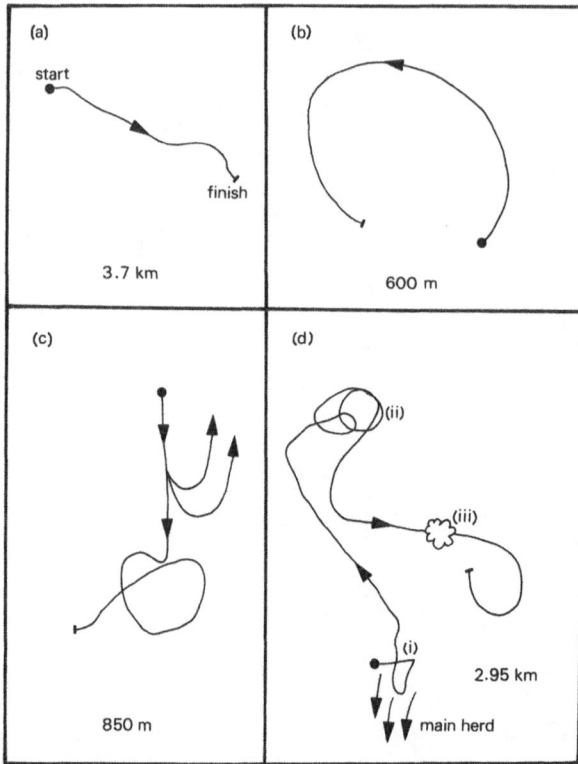

Figure 3.16 Four examples of successful spotted hyaena hunts of gemsbok calves. (a) A single hyaena separated a calf from a herd of about 50 gemsbok. After 2.9 km the calf began to circle slightly and was caught 0.8 km further on. (b) Three hyaenas ran into a herd of about 20 gemsbok, separated a small calf which ran round in a large circle trying to get back to the herd, but was caught after 600 m. (c) One hyaena encountered three calves which it chased. After 200 m two calves doubled back past the hyaena, which continued to chase the third. This one also tried to double back past the hyaena but was unable to do so. After 200 m the calf ran around in a large circle but was eventually caught by the hyaena after a total chase of 850 m. (d) Four hyaenas separated a four-month-old calf from a herd and: (i) chased it as illustrated for 1.55 km, when the calf ran into a group of six gemsbok; (ii) the hyaenas separated the calf and an adult from the herd, and chased them for 900 m, when the gemsbok backed up next to a bush; and (iii) the two gemsbok broke from the bush and after 200 m stood again – this time the hyaenas managed to separate the calf from the adult and 300 m further on killed it.

hyaenas were not found to be more successful when hunting gemsbok calves than groups of four or less (9).

When several hyaenas encountered a herd of gemsbok with calves, it appeared that the hyaenas hunted independently of one another. On one occasion, discerned from spoor, two hyaenas ran into a herd of about 10 gemsbok with at least three calves. Each hyaena selected a calf, the first

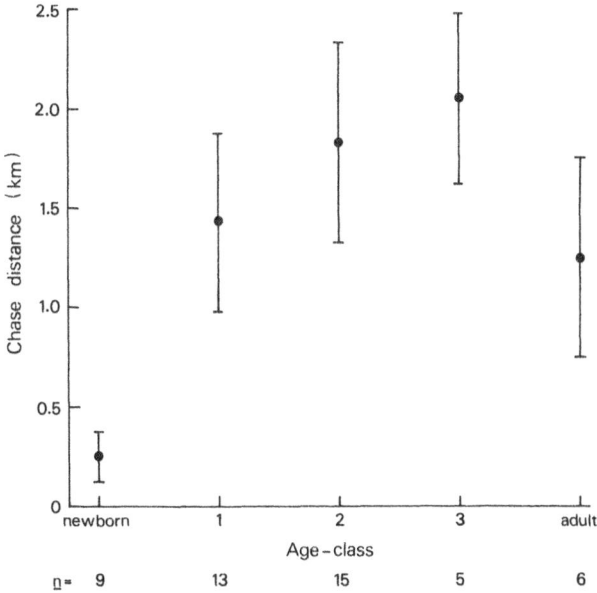

Figure 3.17 Mean ± 2 SE distances which gemsbok of different ages were chased by spotted hyaenas in successful hunts.

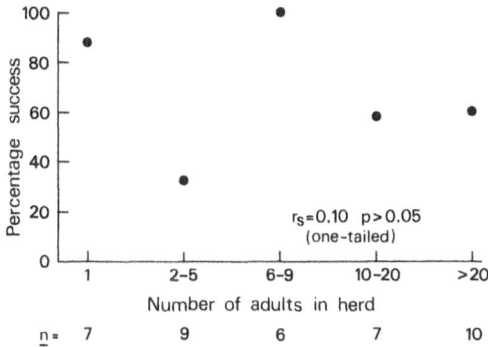

Figure 3.18 The relationship between spotted hyaena hunting success on gemsbok calves and the number of adult gemsbok in the herd.

pulling one down after a chase of 0.45 km, the second killing the other after a chase of 2.2 km.

It seems, therefore, that the most important factor affecting the success with which spotted hyaenas are able to isolate a calf from a herd is the way in which the gemsbok are distributed when the hyaenas first arrive; the more scattered they are the better the hyaenas' chances. Once a calf has been selected and the chase has begun, it has little chance of escaping.

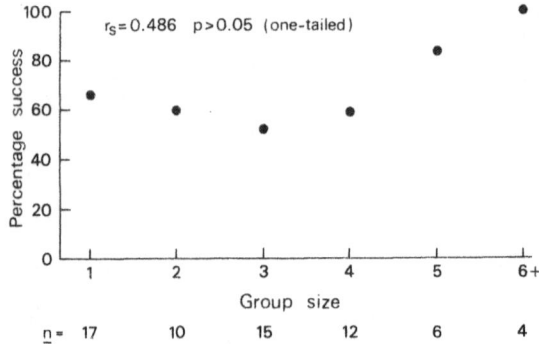

Figure 3.19 The relationship between spotted hyaena hunting success on gemsbok calves and hunting group size.

Occasionally, it may get back to the adults or, if it is old enough, it may be able to outrun the hyaenas.

3.6.2   Gemsbok adults (age-class 3 and above)

Six spotted hyaenas run at an adult gemsbok which runs off for 100 m, then stands and faces them. The hyaenas surround it briefly (Fig. 3.20), then move off . . . six hyaenas run into a herd of five adult gemsbok, the gemsbok run off in different directions, with the hyaenas chasing each gemsbok for a short distance. Then four of the hyaenas run off after another five adult gemsbok which are standing close by. One of the gemsbok puts its head down and chases the hyaenas away . . . seven hyaenas come up to a gemsbok, the gemsbok backs up to a tree, the hyaenas circle it and the gemsbok charges at one with its head down, at which the hyaena runs away. The gemsbok stands, the hyaenas watch it for a few minutes, then move off slowly.

These are examples of typical interactions between spotted hyaenas and adult gemsbok. However, sometimes the interactions are more intense:

> 3 July 1982. Seven Pans. As nine hyaenas come to the crest of a dune at dusk, they stop and look off to their right with their ears cocked, then lope off in that direction. After 500 m I see an adult gemsbok ahead standing on top of a dune 200 m away. As the hyaenas come running up to it the gemsbok runs off. They lope up to the place where it was standing, sniff the ground and run off more quickly after it. After 200 m they come up to two gemsbok, a bull and a cow, which are standing together facing the hyaenas. The hyaenas circle the gemsbok and several dart at the gemsbok, which charge at the attacking hyaenas with their heads held low. A stalemate ensues; hyaenas and gemsbok standing looking at each other.

Figure 3.20   Six spotted hyaenas challenge a gemsbok cow.

Suddenly the cow runs off and is immediately chased by the hyaenas. After 500 m she goes into a large clump of candle acacia, a round bush-clump about 1.5 m high and 10 m in diameter. The hyaenas make some determined efforts to bite the gemsbok in the rear, but she retaliates by swinging her horns at them. Additionally the hyaenas are hindered by the bush. After a few minutes the hyaenas retreat a few metres, some lie down while others stand watching the gemsbok (Fig. 3.21a).

After 24 min the gemsbok slowly moves out of the bush, backs away a few metres, then turns around and runs off. The hyaenas pursue her. She runs for 1 km before taking refuge in another candle acacia bush-clump, as the hyaenas try hard to get at her. The gemsbok's tail has been bitten off by the hyaenas a few centimetres from the base (Fig. 3.21b). The hyaenas are again unable to penetrate the thick bush and soon move away slightly, some lying down, others standing, while three move back 30 m to where the gembok's tail is lying and eat it. The hyaenas remain for another hour, while the gemsbok stays in the bush-clump, but then leave.

On 145 occasions spotted hyaenas were observed in interactions with sub-adult and adult gemsbok (it was impossible during these interactions to differentiate between the two) (Fig. 3.22). Sixty-six of these interactions were not regarded as actual hunting attempts, therefore 11 of 79 (14%) were successful, a much lower success rate than with gemsbok calves. On

(a)

(b)

Figure 3.21 (a) A gemsbok cow takes refuge in a candle acacia bush clump. (b) It does so again after the hyaenas have bitten off its tail.

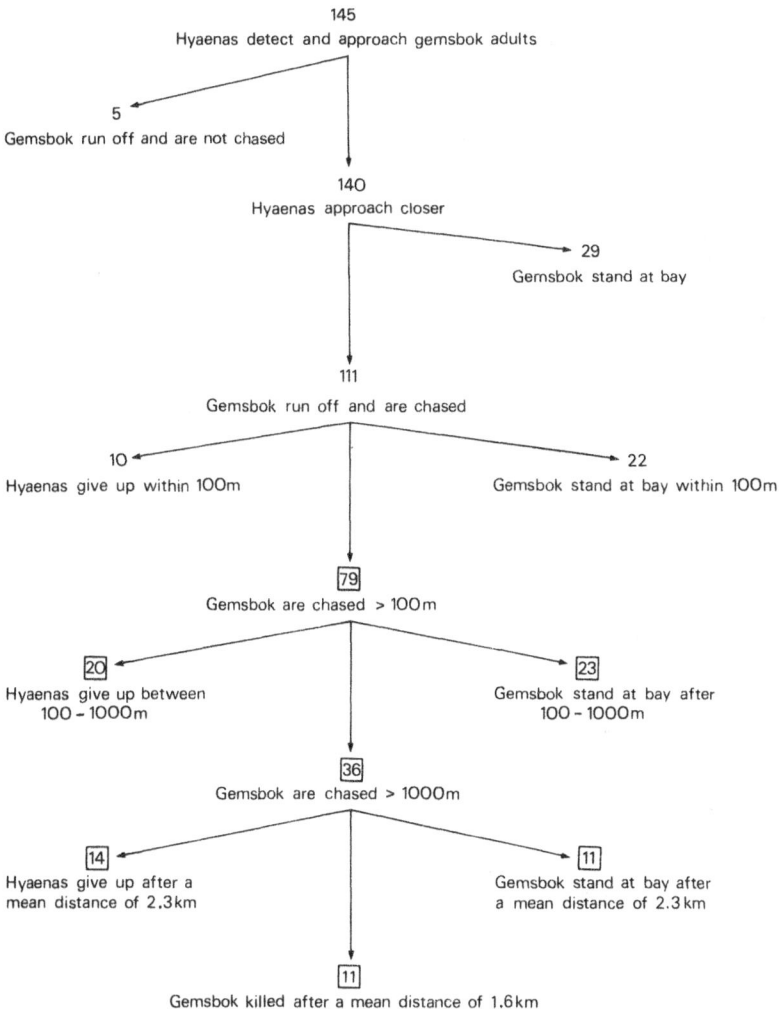

Figure 3.22 Results of interactions between spotted hyaenas and adult gemsbok (boxed figures as in Fig. 3.15).

20 (13.8%) occasions the gemsbok made use of a tree or bush to protect its hindquarters when standing at bay to face hyaenas, and on four (2.8%) occasions it went into a candle acacia clump. On one of these occasions the hyaenas waited 5 h 12 min before finally moving off.

In the 34 cases that the gemsbok ran off for more than 100 m before standing and facing the hyaenas (Fig. 3.22), I judged the effort the hyaenas made to overcome the prey and measured the distance the gemsbok ran. The gemsbok ran further (mean 1.8 ± 0.3 km) when the hyaenas made a determined effort than when the hyaenas were judged not to try hard (mean 0.9 ± 0.3 km) (10).

The group sizes of hyaenas involved in interactions with sub-adult or adult gemsbok were generally higher than those involved in interactions with gemsbok calves (Table 3.4). However, there was no relationship between hunting success of gemsbok sub-adults and adults and the number of hyaenas in the hunting group (Fig. 3.23a), nor between the number of hyaenas hunting and the success rate, if only those confrontations which were regarded as determined hunting efforts (as defined above) are considered (Fig. 3.23b). However, if the number of hyaenas which were involved in successful hunts of sub-adults (age-class 3) and adults (age-class 4 and above) are separated (Fig. 3.24), there was a tendency for larger groups to be involved in kills of adults than sub-adults (11). The smallest group of hyaenas which was observed to kill an adult gemsbok was four, and a single hyaena was never observed to kill a sub-adult or adult gemsbok.

The hyaenas sometimes expended considerable amounts of energy in unsuccessful adult gemsbok hunts, either by chasing a gemsbok for a long distance, or when it stood and faced them, by making a determined effort to overcome it (also a dangerous pursuit because of the lethal gemsbok horns). However, in nine of the 11 cases the hyaenas managed to kill a

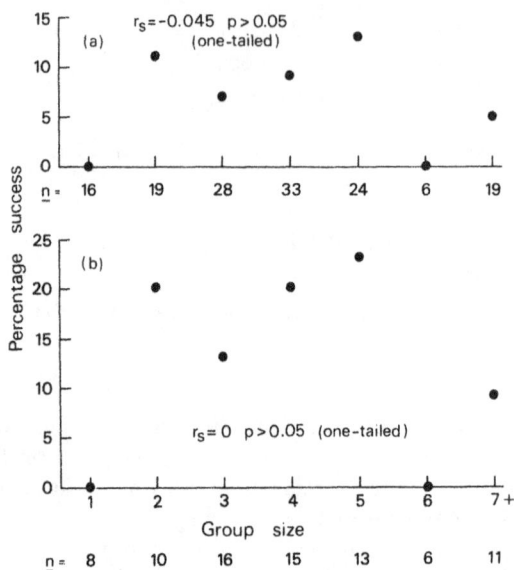

Figure 3.23 The relationship between spotted hyaena hunting success on gemsbok adults and hunting group size: (a) all attempts; (b) determined efforts.

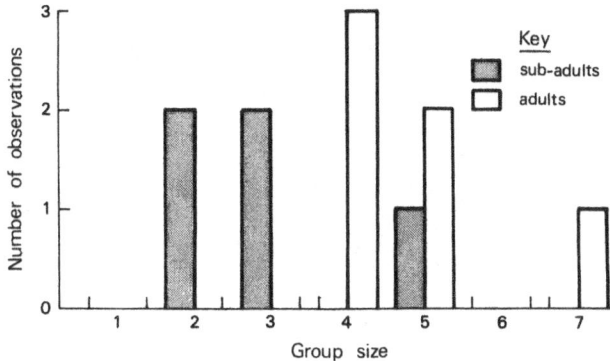

Figure 3.24 Frequency with which various group sizes of spotted hyaenas were observed in successful hunts of sub-adult and adult gemsbok.

sub-adult or adult gemsbok; they did so with relative ease as is illustrated in the following example:

4 April 1982. 02.20 h. Five spotted hyaenas are moving through the dunes near Seven Pans when they stop and stand alert, turning in different directions. After 5 min they run off, obviously having heard something, and 1 km further on they encounter two adult gemsbok. As the hyaenas come running up to them one of the gemsbok runs off 20 m, stops next to a bush and faces the hyaenas. They immediately leave it and run over to the second gemsbok which runs off. The hyaenas give chase at a speed of approximately 30 km/h. After 1.5 km three of the hyaenas catch up to the gemsbok and quickly pull it down onto its side (Fig. 3.25), immediately being joined by the other two. The gemsbok lifts its head a few times and attempts to swing its horns at the hyaenas which are disembowelling it, but the fact that the gemsbok is lying on its side greatly reduces the effectiveness of its horns in defence. Three minutes after being pulled down, the gemsbok, a cow, is dead.

Two cases where an intense struggle ensued before the gemsbok was finally killed were discerned from spoor. In one, five hyaenas struggled with the gemsbok over the last 200 m of a 2.2 km chase, and in the other four hyaenas caught up with the gemsbok after 700 m and pulled its intestines out, but it struggled on for another 20 m until they finally killed it.

The marrow content of the few long-bones examined of adult gemsbok killed by hyaenas (Fig. 2.16), provided some evidence that the hyaenas were selecting animals in poor condition. The way in which hyaenas approach and examine gemsbok and then quickly leave many, but seriously pursue others, suggests that the hyaenas are testing the gemsbok, only

Figure 3.25   Three out of a group of five spotted hyaenas have pulled a gemsbok cow onto its side so that its horns become ineffective, and one of them is disembowelling the gemsbok.

pursuing those that show some indication of being unfit. If this is correct, it is a subtle phenomenon, as, except for one instance of a sub-adult with a twisted horn, I could discern no differences between those gemsbok that were immediately left, those that were hunted unsuccessfully with determination, and those that were killed. Furthermore, the comparatively short distances adults were chased before being caught (Fig. 3.17), also suggests that these individuals may have been unfit. Possibly, the behaviour of the gemsbok determines the effort that the hyaenas put into the hunt. A normal, fit gemsbok, may stand and face a group of hyaenas, whereas a less fit animal does not have the confidence to challenge them. Therefore, if a gemsbok runs the hyaenas pursue it, and the further it runs before standing the harder the hyaenas will try to kill it. A problem with this hypothesis, however, is that it might be possible for the gemsbok to cheat.

The interactions between spotted hyaenas and large gemsbok are similar to those between wolves and moose (Mech 1970, Peterson 1977). The moose often stand and ward off the wolves and only a low percentage, 3.9% of all interactions between moose and wolves in both studies combined, lead to the immediate death of a moose, compared to 7.6% of those between gemsbok and spotted hyaenas. In the moose–wolf studies it has been clearly shown that wolves select old and physically disabled adults.

3.6.3   Wildebeest

25 June 1983. Langklaas. 20.30 h. Four hyaenas come running up to a herd of 50–60 wildebeest. The wildebeest run off in a wide arc bunched together, circling around, the hyaenas pursuing them at a speed of about 20 km/h. After 1 km the wildebeest straighten out as the hyaenas put on the pressure. Two wildebeest break off from the pack, but the hyaenas ignore them. After running for 2.9 km the hyaenas stop and the wildebeest continue running until they are out of sight.

The hyaenas lie down for half an hour, during which time they and others close by whoop several times, and are joined by four other adults and two large cubs. Then they move over to the same herd of wildebeest and chase them as before. The hyaenas soon split up, the one that I follow runs hard after a group of three wildebeest that have broken away from the main herd. The hyaena pushes them and two peel off, but the hyaena continues after the third, a six-month-old calf (I was unable to age the other two). The hyaena pursues the calf for 2.7 km at a speed of 40–50 km/h. For most of the time the calf is far ahead of the hyaena, but towards the end it starts to circle and the hyaena gains on it by cutting corners. As the hyaena closes in, the wildebeest doubles back on its tracks and the hyaena catches it easily, bringing it down and biting at its stomach. Within 30 s a second hyaena comes running up and the two quickly dispatch the calf. Within minutes the other hyaenas have arrived at the carcass, running up to it in ones and twos.

It was not always possible to determine if hyaenas were hunting adult wildebeest or calves – for example, once a group of seven hyaenas ran at a herd of wildebeest, four brought down an adult cow after a chase of 500 m, while one killed a six-month-old calf after 2.2 km. Accordingly, the analyses of spotted hyaenas hunting wildebeest have been split between those of hyaenas hunting mixed (adults and calves) herds (Fig. 3.26a), and those of hyaenas hunting adult bull herds (Fig. 3.26b).

Combining the data from Figures 3.26 a and b and discarding eight interactions between hyaenas and wildebeest bulls where the bulls either stood at bay, or the hyaenas gave up within 100 m, 16 of 41 (39%) hunting attempts on wildebeest were successful. Kruuk (1972) recorded an overall hunting success rate by hyaenas on wildebeest of all ages of 34%. Given the difficulties in comparing these data (section 3.5), the hunting success rate of spotted hyaenas on wildebeest in the Kalahari and in East Africa appears to be of the same order of magnitude.

There was no obvious relationship between the distance of the chase and the age of the victim (Fig. 3.27). The mean distance of successful wildebeest hunts (1.0 ± 0.2 km, range 0.15–2.7 km), however, was shorter than the mean distance of unsuccessful chases (2.1 ± 0.5 km, range 0.5–4.2 km) (12). Outrunning the hyaenas, therefore, was the most common way in which a selected individual wildebeest man-

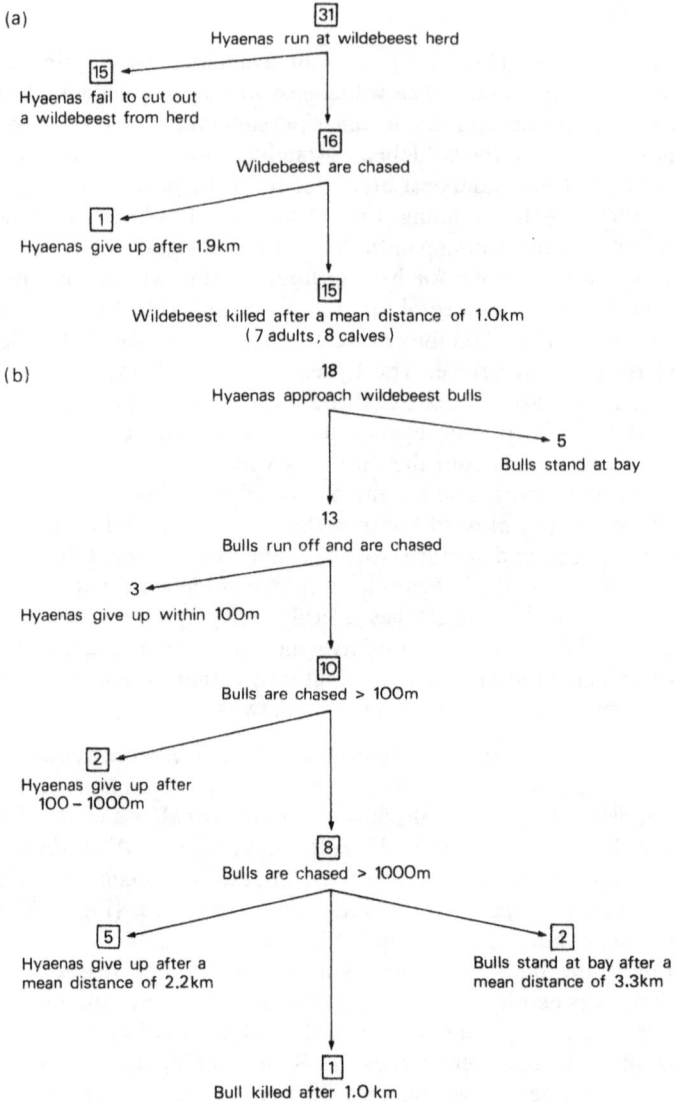

Figure 3.26 Results of interactions between spotted hyaenas and (a) mixed wildebeest herds of 20 or more animals, and (b) bull herds (boxed figures as in Fig. 3.15).

aged to escape being killed. This is in agreement with Kruuk's (1972) observations.

Although, as also recorded by Kruuk (1972), there was usually little active defence by a wildebeest once it had been caught up with, on two occasions an adult wildebeest managed to defend itself successfully. On

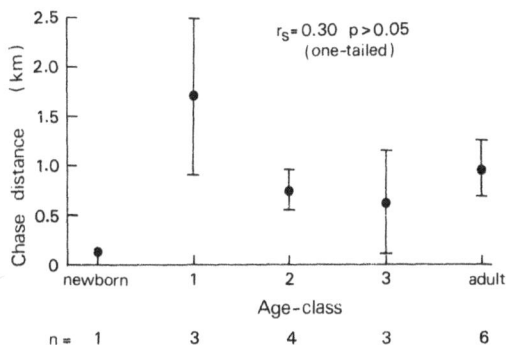

Figure 3.27 Mean ± 2 SE distances which wildebeest of different ages were chased by spotted hyaenas in successful hunts.

one, which was discerned from spoor, three hyaenas chased a lone wildebeest for 2.8 km before it stood, and, judging by the disturbance in the sand, the hyaenas had made a determined, but unsuccessful attack on it. They then chased it for 1.4 km before leaving it. On the other occasion a wildebeest bull backed up against a fibreglass drinking reservoir and kept three hyaenas at bay for 7 min, once nearly crushing a hyaena between itself and the reservoir. A third observation, not incuded in Figure 3.26, was when I came across seven hyaenas which were attacking a wildebeest bull in a large rain pool in the road. The wildebeest swung its horns at the hyaenas, once actually catching one with a horn and throwing it a metre into the air. It took the hyaenas over 25 min to kill the wildebeest.

The habit of wildebeest running into water when chased by spotted hyaenas has also been recorded by Kruuk (1972). Besides the observation described above, another three occasions were documented of a wildebeest running into one of the large drinking reservoirs at a windmill and being killed by hyaenas. As concluded by Kruuk (1972), this anti-predatory behaviour does not seem to have survival value against hyaenas. Perhaps it is a more successful anti-predatory defence against other cursorial hunters such as wild dogs.

Figure 3.28 shows the relationship between spotted hyaena hunting success on wildebeest and hunting group size. As with gemsbok there does not appear to be a relationship. Although groups of seven or more hyaenas were more successful than were groups of six or less (13), there was no difference in hunting success between groups of 1–3 versus groups of 4–6 hyaenas (14). The apparent sharp increase in hunting success between groups of six and groups of seven or more hyaenas is probably a result of small sample sizes rather than a biological phenomenon.

In conclusion, the hunting behaviour of spotted hyaenas towards wildebeest in the southern Kalahari is similar to this behaviour in East Africa, except that in the Kalahari there is less pressure exerted by hyaenas on

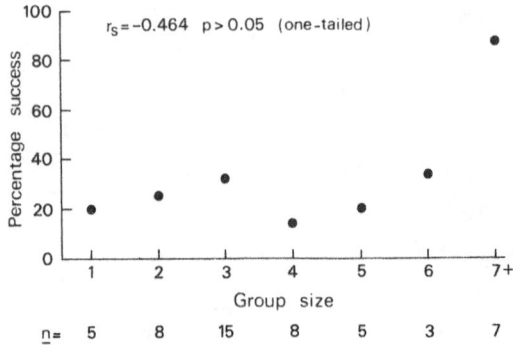

Figure 3.28 The relationship between spotted hyaena hunting success on wildebeest and hunting group size.

young calves. Their hunting technique suggests that hyaenas select their prey visually, although the small sample of bone marrow from kills of adults (Fig. 2.11) could not show selection for adult animals in poor condition.

3.6.4 Springbok

Springbok, the most abundant ungulate along the river beds (Fig. 2.3), were usually ignored by spotted hyaenas. The exception was during January and February, usually the peak springbok lambing season (section 2.1.1.2), when there was a significant increase in hunting attempts on springbok (Fig. 3.29) (15). Most hunting attempts were on lambs and were similar to the brown hyaena's behaviour in similar conditions (section 3.5.1), and to spotted hyaenas hunting Thomson's gazelle in East Africa (Kruuk 1972).

Because it was usually difficult to separate hunting orientated towards lambs from that orientated towards adults the observations have been combined (Fig. 3.30). Of the 41 interactions between spotted hyaenas and springbok, 13 were regarded as definite hunting attempts, when the hyaenas were judged to have chased the springbok as hard as they could, of which 4 (31%) were successful.

Chase distances were short. Once, two hyaenas caught a lamb after an 800 m chase, and a sub-adult was caught in 30 m. The mean distance of unsuccessful chases was 644 ± 198 m, the longest being 1700 m. Small lambs usually lie still when approached by a hyaena (spotted or brown), and several times a hyaena passed within 2–3 m of a lamb, without detecting it. Twice, a spotted hyaena came across a lamb and snapped it up.

Spotted hyaenas usually hunted springbok solitarily (Fig. 3.6). Three of the kills were made by a single hyaena and the fourth by two. The hyaenas

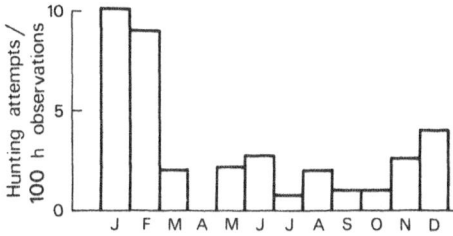

Figure 3.29 Hunting attempts per 100 h observations by spotted hyaenas on springbok each month of the year.

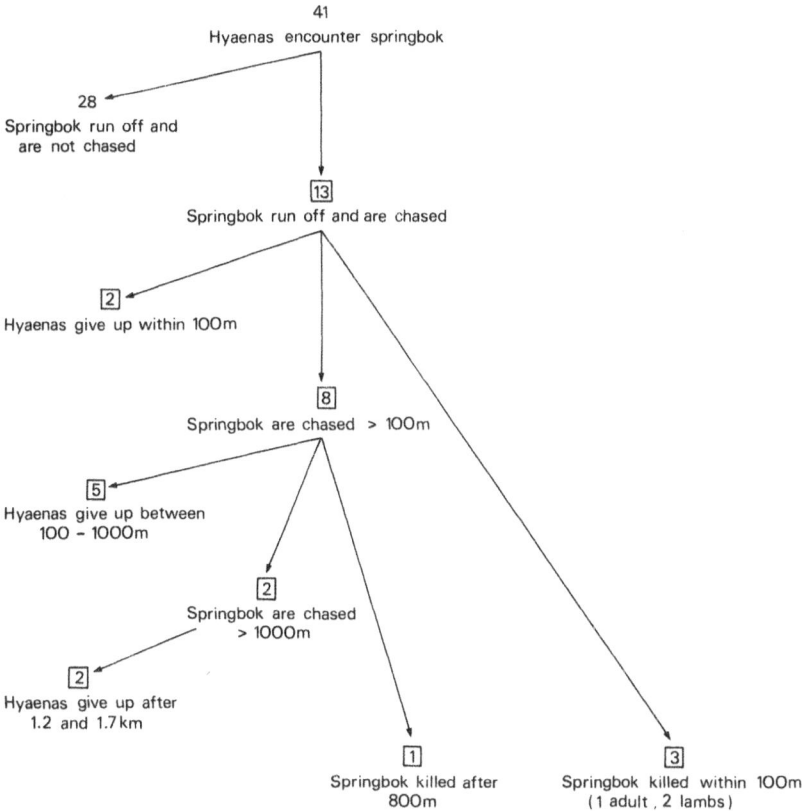

Figure 3.30 Results of interactions between spotted hyaenas and springbok (boxed figures as in Fig. 3.15).

appeared to run as fast as they could when hunting springbok, but adult springbok did not appear to extend themselves when being chased, easily keeping in front of the hyaenas. In eight (20%) of the 41 interactions the springbok were observed to 'pronk' – progress by leaping off the ground with all four legs held stiff and straight – (Bigalke 1972) as the hyaenas ran after them, a behaviour similar to stotting by Thomson's gazelle (see Walther 1969, Caro 1986 a, b, FitzGibbon & Fanshawe 1988).

Adult springbok are normally too fast for spotted hyaenas and the hyaenas waste little energy in futile attempts to catch them. Lambs are more easily caught and, as discussed in section 2.3.3.4, are probably more frequently caught than the data show.

3.6.5   Eland

Eland were only sporadically available to hyaenas in the study area, but when they were, calves in particular appeared to be an important prey species (Table 2.15).

Of the 14 interactions between eland and spotted hyaenas that were judged to be hunting attempts, seven (50%) were successful (Fig. 3.31). On two occasions the hyaenas managed to kill two calves at a time. Once, nine hyaenas were walking along the Nossob river bed when three young eland calves came running towards them – the hyaenas killed two of the calves (Fig. 3.32), but the third somehow escaped. On another occasion

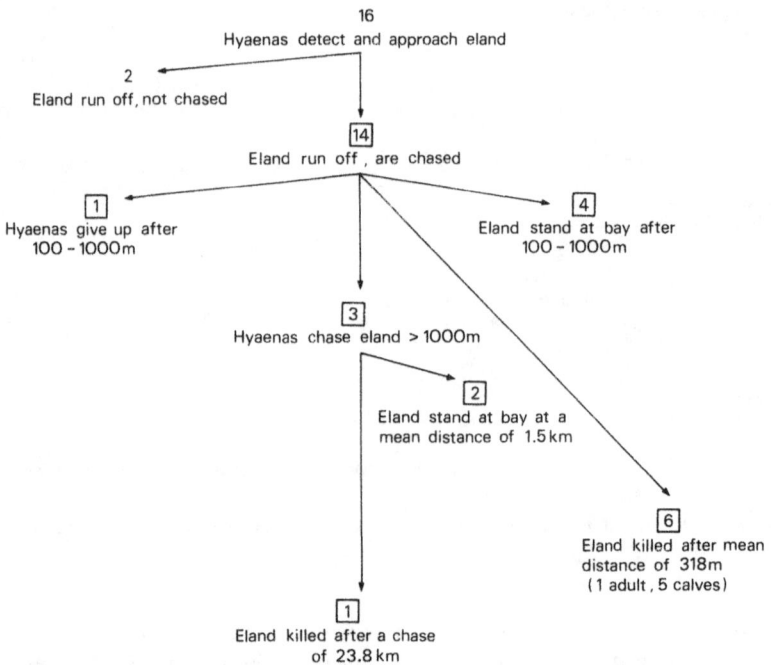

Figure 3.31   Results of interactions between spotted hyaenas and eland (boxed figures as in Fig. 3.15).

Figure 3.32 Nine spotted hyaenas feed on two eland calves they have killed.

(discerned from spoor), two hyaenas ran into a herd of eland and each brought down a calf after chases of 300 m and 400 m.

The distances of successful hunts were: newborn calf 10 m ($N = 1$); calves mean $= 280 \pm 37$ m ($N = 4$); a sub-adult 23.8 km; an adult 900 m. The mean distance hyaenas ran after eland in unsuccessful hunts was 633 ± 145 m ($N = 6$), except for one case when they followed the eland for 11.1 km. Once, the hyaenas gave up the chase after 0.3 km, but in the remaining five instances the eland, ranging in number from 1 to 15, defended themselves either by swinging their horns at the hyaenas, or by standing in a circle with their heads together and kicking out vigorously at them with their hind legs.

Only one adult eland kill was seen. Three hyaenas located a cow on its own after following a scent trail for 1.3 km. As they ran up to it, the eland defended itself for a few seconds, then ran off. It was caught after a chase of 0.9 km, and killed within 5 min, putting up little resistance. While the hyaenas were feeding a small calf appeared, which a hyaena ran up to and killed.

The data from eland hunts are too few to statistically test the relationship between hunting success and group size, although this does not appear to be strong (Table 3.8).

The tenacity and opportunistic nature of the spotted hyaena's hunting behaviour is illustrated in the case of three hyaenas following a herd ·of about 20 adult eland for 23.8 km. Eventually, after challenging the eland several times, they managed to get hold of a sub-adult bull and, after a

Table 3.8   Group size and rates of success of spotted hyaenas hunting eland.

	Hyaena group size								
	1	2	3	4	5	6	7	8	9
Successful hunts									
Adults	0	0	2	2	0	0	0	0	0
Calves	1	1	1	0	0	0	0	0	1
Unsuccessful hunts	0	0	1	3	2	0	3	0	0
Total	1	1	4	5	2	0	3	0	1

struggle which stretched over 1 km, kill it. The following is a similar observation, although in this case the hyaenas were unsuccessful:

28 January 1980. 03.43 h. Dunes west of Kousaunt. Three spotted hyaenas run off after a young adult eland bull. After 750 m it stops, and the hyaenas attack it, trying to bite it at the shoulder and rear, but the eland wards them off with its horns. It runs off again and the hyaenas chase it at a speed of approximately 20 km/h, repeatedly darting in, then retreating, trying, it appears, to hamstring the eland. After 1 km the eland stops again and defends itself by swinging its horns and kicking out vigorously with its back legs. For the next 10 min the eland moves back and forth over an area of some 400 m followed by the hyaenas, stopping intermittently and defending itself, as the hyaenas attack it. Once, one of the hyaenas manages to bite it in the shoulder, but it does not draw blood.

By 03.58 h the attacks are getting less intense and the hyaenas are merely harassing the eland, following it when it moves, standing, or even lying down, when it stands. When moving with the eland there is one hyaena out in front and two behind, with one of the latter right on its heels. After following it for 11.1 km, 2 h 8 min from the time that they first encountered it, the hyaenas finally leave the eland at 06.00 h.

There are other examples of spotted hyaenas attacking very large prey. Kruuk (1972), observed attacks on black rhinoceros and their calves, and buffalo. Henschel (1986), observed spotted hyaenas successfully kill a hippopotamus, and twice attack a giraffe cow and calf, and A. Starfield (personal communication) also observed hyaenas kill a hippopotamus. Hitchins & Anderson (1983), provided circumstantial evidence for spotted hyaenas frequently attacking black rhinoceros calves in Natal. The chances of success against such large prey are limited, the attacks are fraught with dangers to the hyaenas (Starfield saw a hippopotamus crush a hyaena's head between its jaws), and considerable energy is expended, although the potential rewards are high.

3.6.6   Other ungulates

3.6.6.1   HARTEBEEST

Spotted hyaenas showed interest in hartebeest on only 10 occasions. Twice, hyaenas ran at a herd of hartebeest which ran off and were not pursued, and seven times, hyaenas ran after a single hartebeest for a mean distance of 457 ± 245 m, before losing interest. The longest of these chases was 1.9 km, and the two hyaenas appeared to be running as fast as they could, whereas the hartebeest appeared not to extend itself. It seems that hartebeest are normally too fast for hyaenas. The only successful hunt of a hartebeest I saw was when eight hyaenas easily brought down a cow handicapped by sarcoptic mange.

3.6.6.2   KUDU

The only kudu hunted were two bulls which were killed by the same hyaenas in the space of four nights. From spoor it was discerned that at least three hyaenas had chased the first bull for more than 1 km. In the second case, ten hyaenas chased independently after several kudu. The two hyaenas that I was following stopped after 0.9 km, turned round and then ran off in a different direction for 1.5 km, to where four others had pulled down a bull.

3.6.6.3   STEENBOK

Only six instances of spotted hyaenas hunting the common and widespread steenbok (Tables 2.2 and 2.4) were recorded, and in all cases a single hyaena was involved. Hyaenas chased steenbok for 50, 300 and 500 m before giving up, and once a steenbok escaped into an aardvark hole after running for 250 m. Two hunts were successful: a lamb was flushed and caught after 5 m, an adult was chased for 100 m, caught but escaped, and was caught again and killed 150 m further on.

3.6.7   Non-ungulate mammals

Spotted hyaenas occasionally hunted some non-ungulate species:

3.6.7.1   PORCUPINE

Twelve interactions between spotted hyaenas and porcupines were observed, involving between one and nine hyaenas. Nine times the hyaenas came up to the porcupine, looked or sniffed at it briefly, or followed it a short distance, and then lost interest. On two other occasions the porcupine took refuge in a hole. Once, one of nine hyaenas killed a porcupine, but no details were gleaned. Spotted hyaenas frequently use porcupine holes as dens (section 6.1), and in two instances killed a porcupine shortly after moving in.

3.6.7.2 MICE
A large spotted hyaena cub chased a mouse which crossed its path and caught it after several attempts. Twice, adult hyaenas bit at mice, but did not kill them.

3.6.7.3 OTHER RODENTS AND HARES
Five times, once with success, single-spotted hyaenas chased a springhare for distances varying between 100 and 300 m. Once, a spotted hyaena moving in a group of eight flushed and caught a hare leveret. Twice, single hyaenas, and once, three, chased an adult hare for between 5 m and 400 m, in the latter case with success.

3.6.7.4 CANIDS
On five occasions single-spotted hyaenas chased a bat-eared fox. On one occasion the hyaena gave up after 200 m, and on two others the fox escaped into a hole. On the fourth occasion a hyaena chased a bat-eared fox for 500 m, during which the fox stopped suddenly three times to face the hyaena, with its tail in the inverted U position (Kleiman 1967), and uttering a rattling growl (see section 3.5.1.3). The hyaena made no attempt to attack the fox and appeared to get a fright each time the fox turned and faced it. The final observation was when a hyaena flushed a bat-eared fox and chased it for 200 m, before the fox doubled back past the hyaena and ran straight into my vehicle! The hyaena ran up to the stunned fox and killed it, but did not eat it. A black-backed jackal cub approached a sub-adult spotted hyaena, which chased it several times round in a circle of about 10 m in diameter, biting the jackal three times. Each time the jackal was bitten it yelped. The jackal then ran over to a dead tree, round which the hyaena again chased it several times, until the jackal ran off.

3.6.7.5 MUSTELIDS
Spotted hyaenas were seen in three interactions with honey badgers. In one, three hyaenas ran after a honey badger which disappeared over a rise. In another, a honey badger turned and faced three hyaenas, emitting the grating, rattling sound referred to in section 3.5.1.5, at which the hyaenas left it alone. In the third case four hyaenas chased a honey badger which took refuge in a tree. However, it immediately fell out of the tree amongst the hyaenas, and after a scramble ran off followed by the hyaenas. As they caught up with it, it stood and faced them rattling, and emitted a foul smell. This checked the hyaenas and the honey badger ran off – a few metres further on the same thing happened again. One hyaena continued to follow the honey badger, but after 50 m it also gave up.

Two sub-adult hyaenas chased a striped polecat for 200 m. The polecat ran away and also emitted a smell. Several times, one of the hyaenas made as if to bite it, but drew back quickly. Once, one of them picked the polecat up and tossed it into the air a metre or so. The polecat finally escaped into a hole.

3.6.7.6 AARDVARK
Six instances of one to five spotted hyaenas hunting an aardvark were recorded. On all occasions the aardvark escaped into a hole after chases varying from 20 to 400 m.

3.6.8   Ostrich

5 December 1979. 23.30 h. Five spotted hyaenas foraging in the dunes 15 km east of Bedinkt start moving quickly upwind, and 400 m further on encounter a pair of ostriches with about ten quarter-grown chicks. The ostriches scatter in all directions chased by individual hyaenas. A hyaena bites a chick, leaves it still alive and chases after another. A second hyaena runs past the still fluttering chick and off in the direction the other had taken. However, it soon comes back to the chick and kills it. I then see another hyaena carrying a second dead chick. Hyaenas are running back and forth with noses to the ground, obviously still looking for chicks, but they have escaped.

It seemed that the hyaenas had tried to catch as many chicks as possible before they escaped, and did not waste time killing and eating the chick that had been immobilized. The ostriches appeared to confuse the hyaenas by running off in different directions. This was the only time that I observed spotted hyaenas encounter ostrich chicks.

They frequently encountered ostrich adults, but usually ignored them. Only once did hyaenas chase adult ostriches, apart from the observations at nests described in section 3.4.2. In this instance, two hyaenas ran after five ostriches for 200 m before giving up. Spotted hyaenas once killed an ostrich, which was in a weak state due to the vertebral bone of a large herbivore stuck in its throat (A. Putter personal communication).

3.6.9   Clan differences in hunting success

Hunting success of spotted hyaenas from different clans varied significantly (Fig. 3.33) (16), but this did not appear to be related to differences in hunting ability. For example, the success rate of hyaenas from the Kousaunt clan, varied from 57% in the 1981–1982 period (KS1), to 43% in the 1979–1980 period (KS2), to 92% in the 1983 period (KS3), when essentially the same hyaenas were involved (Fig. 4.6). At least one contributory factor appeared to be prey selection, as there was a negative correlation between hunting success and the frequency with which serious

attacks on adult gemsbok were made (Fig. 3.34); i.e. less successful
hyaenas were more often tackling harder to kill prey.

Why should some hyaenas attack more difficult prey than others? In the
St John's clan's territory, where hunting success was the lowest and serious
attacks on adult gemsbok the highest (Fig. 3.34), the dearth of large
ungulates in the area during the study (Table 4.6) may have forced the
hyaenas to take chances with more dangerous prey. Moreover, although I

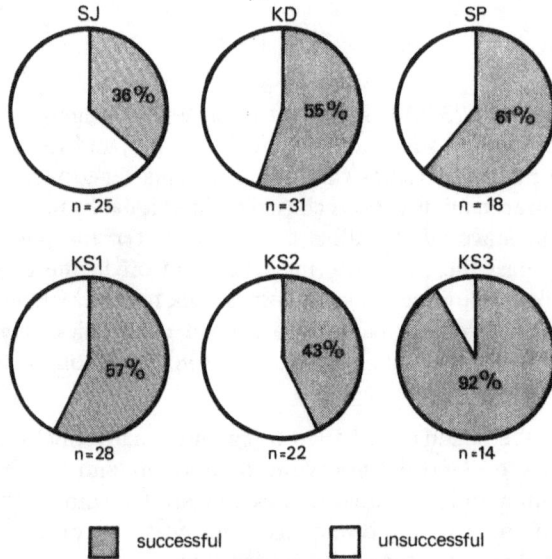

Figure 3.33 Hunting success rates of spotted hyaenas from different clans. Abbreviations for
clans as in Figure 4.21.

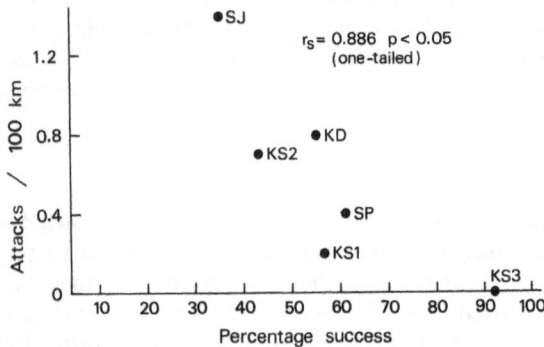

Figure 3.34 The relationship between hunting success on gemsbok adults and the frequency
with which they were hunted by spotted hyaenas from different clans. Abbreviations for clans
as in Figure 4.21.

was unable to quantify it, the intensity of the attacks during unsuccessful hunts on adult gemsbok often appeared to be higher with the St John's clan hyaenas than with others. It may also be relevant that the phenomenally long chase of 23.8 km after the eland was made by hyaenas from the St John's clan.

In other territories this trend was not apparent. The Kaspersdraai and Kousaunt 2 territories had comparatively high densities of large ungulates (Table 4.6), as well as comparatively high incidences of attacks on adult gemsbok (Fig. 3.34), although the figures for these territories were much less extreme than the St John's figures.

3.6.10   Some evolutionary considerations of the spotted hyaena's hunting behaviour and the anti-predator strategies of gemsbok

Like their counterparts in East Africa (Kruuk 1972), spotted hyaenas in the southern Kalahari are efficient hunters. Although they hunt a rather different spectrum of prey, their basic hunting patterns are similar – they select the physically weakest prey, particularly the young. But unlike hyaenas in East Africa, which predominantly take wildebeest calves under six months of age (Kruuk 1972), southern Kalahari hyaenas also regularly hunt larger calves of six to 12 months of age of both wildebeest and gemsbok.

In terms of the amount of food they provide per hyaena, eland, gemsbok, and wildebeest adults are the most rewarding species for hyaenas to hunt in the southern Kalahari (Table 3.9). However, eland adults because of their size, and gemsbok adults because of their horns, are not easy to kill. Wildebeest adults on the other hand may be easier to kill.

Table 3.9   The approximate amount of food available per spotted hyaena from various types of killed prey. Mass figures are taken from Smithers (1971) and Meissner (1982). For large mammals 33%, and for medium-sized 10%, of carcass weight has been subtracted for non-edible parts.

Carcass	Number of observations	Mean mass (kg)	Mean number hyaenas eating	Mass available/ hyaena (kg)
Gemsbok adult	13	132	6.2	21.3
Gemsbok calf	42	45	4.5	10.0
Wildebeest adult	10	132	6.9	19.1
Wildebeest calf	10	45	4.9	9.2
Springbok lamb	5	5	1.2	4.2
Eland adult	3	330	6.3	52.4
Eland calf	7	72	4.6	15.7
Hartebeest adult	3	96	8.7	11.0

Eland calves appear to be more rewarding to kill than gemsbok calves (Table 3.9), and the impression was that when eland calves come into an area, they were hunted in preference to gemsbok calves. Springbok lambs are a relatively low-yielding prey species (Table 3.9), and are not hunted with fervour by spotted hyaenas.

Gemsbok calves up to one year of age are the most frequently killed animals by southern Kalahari spotted hyaenas, and their success rate at catching these prey is relatively high (63%). Although they are present throughout the year, the distribution and numbers of gemsbok calves is such that the hyaenas are often forced to cover large distances before encountering them.

Adult gemsbok can normally defend themselves against spotted hyaenas with their long rapier-like horns, which are well developed in both sexes. The shape of the horns, the fact that there is little sexual dimorphism in the horns and their use in defence against spotted hyaenas and lions (Eloff 1964), suggest that the main evolutionary pressure on horn development in gemsbok was for defence against predators (see Packer 1983). Gemsbok do not have the speed of hartebeest and springbok, nor the stamina of wildebeest. Moreover, the comparatively rapid rate at which their horns grow in the first year (Table 3.10), also suggests that defence is an important function of gemsbok horns. Once gemsbok are one year old and their horns are approximately three-quarters of the adult's length, spotted hyaenas have difficulty in overpowering them.

A remarkable feature of spotted hyaena hunting behaviour, noted by Kruuk (1972) in East Africa, was the way in which they hunted different prey species in different sized groups, and how they often apparently 'preselected' the species they were going to hunt. For example, they set out in larger foraging groups when hunting zebra than when hunting wilde-

Table 3.10   Approximate percentage of adult horn length at different ages in various antelope species.

Species	Approximate percentage of adult horn length at:			Source
	6 months	12 months	18 months	
Gemsbok	53	74	83	Mills, unpublished data
Eland	45	61	77	Jeffery & Hanks 1981
Sable	35	55	78	Grobler 1980
Springbok	31	42	82	Rautenbach 1971
Grey rhebok	30	51	78	Rowe-Rowe 1973
Thomson's gazelle	13	53	83	Brooks 1961
Impala	12	38	63	Spinage 1971
Bushbuck	11	30	66	Simpson 1972
Kudu	5	14	39	Simpson 1972
Defassa waterbuck	0	11	23	Spinage 1967

beest. The variations in group size of spotted hyaenas hunting different prey species observed in the present study (Fig. 3.6), and the fact that they often set out at night in these group sizes is suggestive of a similar phenomenon.

Why do hyaenas form different sized groups when hunting different prey species? Cooperation in overcoming large and fast prey, where group defence of calves is often employed, would seem to be an important factor, and is often cited as such in group-hunting carnivores (Mech 1970, Kruuk 1972, Ewer 1973). However, group size and hunting success were not strongly correlated in the present study (Figs 3.19, 3.23, 3.28 and Table 3.8), which would be expected if group hunting had evolved solely to improve hunting success. While there is a certain benefit to group hunting in overcoming prey like adult gemsbok (the smallest group observed to do so was four), hunting groups often deviated from the optimum for hunting success and food intake per individual, as has also been shown to be the case with lions (Packer 1986). This point is discussed further in section 7.3.3.2.

3.7   Feeding behaviour

The differences in diet and foraging behaviour between the brown hyaena and spotted hyaena mean that the two species feed under different circumstances. The brown hyaena usually feeds alone, the spotted hyaena in a group. The number of hyaenas at a carcass affects their behaviour, especially in the spotted hyaenas' case, so that there are regional differences in feeding behaviour. In this section I contrast some of the differences in feeding behaviour between the two species, and make some intraspecific comparisons between populations of spotted hyaenas.

3.7.1   Behaviour at carcasses

3.7.1.1   BROWN HYAENA

Much of the brown hyaena's food consists of small pieces of bone, odd legs and skulls, etc. which are consumed alone. When feeding on the last remains of a carcass, a brown hyaena will move in ever-increasing circles with its nose to the ground, searching for the last few bone chips and other scraps.

When a large herbivore carcass consisting of skin and bones is found, a brown hyaena has difficulty in breaking off and consuming the large bones, as the following example illustrates:

17 January 1975. Seven Pans. A female brown hyaena finds the fresh skin and bones of a hartebeest carcass at 02.30 h. She begins to chew on the broken head of a tibia, but after 16 min gives up without much

success. She then feeds on the ribs for 5 min, easily cracking and swallowing them, before picking up another leg, which has already been broken off from the carcass. She holds it upright with her forefeet and chews on the proximal head of the tibia, licking it vigorously before biting it, and using her neck muscles in pushing down and sideways, as she struggles to break the bone (Fig. 2.8). After 49 min she finally breaks the head of the bone and starts eating it, moving onto the shaft which she consumes easily. She then picks up a piece of hide, shakes it vigorously and chews on it for 5 min. At 04.07 h, after trying for 12 min, she breaks off a radius and ulna and carries them off.

Ewer (1954) has shown that of the three hyaenas the spotted hyaena has the most specialized dentition, followed by the brown and then the striped hyaena. This applies particularly to the carnassial sheer and to the larger size of the principal bone-crushing tooth, the third premolar in the upper jaw. The above, and other observations, of brown hyaenas struggling to break off and consume large bones, and similar observations of spotted hyaenas which appeared to accomplish this more easily, bear this out.

Carcasses of medium-sized animals were consumed by brown hyaenas far more easily and rapidly than were large ones. For example, the entire skeleton of a springbok, except for the head, was eaten in 1 h 23 min. The brown hyaena was able to deal with the bones easily, beginning with the ribs, which still had a little meat on them, then eating the vertebral column and finally the legs. If there was a substantial amount of meat on a carcass,

Figure 3.35　Four brown hyaenas and two black-backed jackals scavenge a gemsbok.

Figure 3.36   A brown hyaena carries off the leg of a gemsbok as a second one feeds.

a brown hyaena would either break pieces off and carry them away for storage (section 3.7.2.1), or carry some to the den for the cubs (section 7.22).

Several brown hyaenas sometimes accumulate at a large carcass. If there is much meat on the carcass the hyaenas feed together (Fig. 3.35), but normally only one brown hyaena feeds at a time. While one is feeding, the others lie close by waiting for it to vacate the carcass. An individual will usually leave when it has succeeded in breaking off a piece which it carries off (Fig. 3.36). Owens & Owens (1978) recorded similar behaviour in brown hyaenas in the central Kalahari at large carcasses. However, they claimed that the amount of time a brown hyaena spent feeding at a carcass depended on its position in the dominance hierarchy. I did not find this; in fact little evidence for a dominance hierarchy amongst southern Kalahari brown hyaenas was found (section 7.2.3).

3.7.1.2   SPOTTED HYAENA

Kruuk (1972:124) described how 30 or more spotted hyaenas in East Africa often crowded around a carcass so that it became 'completely buried beneath a writhing mass of hyaena bodies'. Competition between feeding hyaenas was expressed in the speed of eating, rather than in actual fighting. In contrast, Tilson & Hamilton (1984) described a situation in the Namib Desert where spotted hyaenas usually fed one at a time, or a female with dependent offspring, and where large carcasses were consumed over several nights, i.e. like brown hyaenas. These differences are ascribed by Tilson & Hamilton to differences in densities of spotted hyaenas and their competitors in the two areas.

Immediately an animal is killed in the southern Kalahari, irrespective of how many hyaenas are present (the most observed was 11), they devour the carcass in a similar manner to the East African situation (Tilson & Hamilton were never present at the kill). Within 2–30 min, depending on the size of the carcass and the number of hyaenas present, the pace slows down (Fig. 3.37), and some hyaenas may leave the carcass to lie down close by. This is at least partially determined by the relative dominance and kinship relations of the hyaenas at the carcass, as will be discussed in sections 7.2.3 and 7.2.4. At scavenged carcasses the hyaenas feed in relatively smaller groups at a slower pace, more in the Namib Desert pattern.

If a carcass is not consumed on the first night the hyaenas will often drag it under a bush and spend the next day lying close by. However, once they abandoned a carcass they were never observed to return to it the next night. On several occasions hyaenas abandoned carcasses which still had a large amount of meat on them (see section 2.3.4). Usually these were adult females with small cubs at a den, but sometimes sub-adults or males would go with the females. In contrast, Namib Desert hyaenas often abandoned a carcass for several hours during the night, and always during the day, when

Figure 3.37 Eight spotted hyaenas feed with equanimity half an hour after killing an adult gemsbok.

they would often move many kilometres back to the den before returning the next evening (Tilson & Hamilton 1984, J. Henschel personal communication).

Kruuk (1972) recorded that when a hyaena killed small prey it invariably kept the food, even though it may have been challenged by other hyaenas. This was not always the case in the present study. Three times a sub-adult killed a small animal in the presence of other, older hyaenas, and lost the kill to one of them.

3.7.2 Food storing

3.7.2.1 BROWN HYAENA

When a brown hyaena finds a large amount of food it will quickly eat some of it, then carry parts off to be stored, before returning to feed on what remains. A spectacular example of this behaviour was when one found an ostrich nest, mentioned in section 3.4.2:

1 August 1975. 20.05 h. Seven Pans. A brown hyaena arrives at a deserted ostrich nest with 26 eggs in it (Fig. 3.9). She sniffs at the eggs for a few seconds, picks one up and carries it off 50 m, puts it down in the open, then runs back to the nest. She then eats two of the eggs and at 20.25 h picks up another and moves off with it, but I do not follow her. After 20 min she returns to the nest, picks up another egg, carries it off in a northeasterly direction for 450 m, and puts it down in the middle of a clump of tall, thick grass. She comes back to the nest and again carries off an egg in a northerly direction for 600 m, before putting it down under a bush and returning to the nest. She carries off another 10 eggs in different directions, for distances varying between 150 m and 600 m, and hides them under various bushes, or in grass clumps. At 23.36 h she eats another egg at the nest, then carries one off for 600 m, drops it rather carelessly in a grass clump and at 00.12 h (just over 4 h after finding the nest) lies down close by. At this stage there are eight uneaten eggs in the nest.

The next afternoon when I return to the nest at 16.45 h there are only two broken eggs remaining. I find the hyaena and a second one close by, both eating eggs, after which the original hyaena picks up an egg stored under a bush and carries it off for 2.1 km, before putting it down in some long grass. She then moves away from the nest area and forages until 02.00 h, when she returns. She finds and eats two more of the stored eggs, picks up a third and carries it off for 0.9 km before again placing it under a bush. She then moves over to a different bush, where there is another egg, eats this and lies down at 03.21 h

A comparable observation was made when following another brown hyaena, when she moved straight to a bush out of which she collected and ate an ostrich egg. She then investigated several bushes in the vicinity, and after 40 min found another egg. She then moved away from the area and 1 km further on passed an old ostrich nest, with a few broken pieces of egg-shell scattered around it. The next night she came back to the same area, and after casting around for a short while found another two stored eggs.

Examples of red foxes 'scatter hoarding' eggs have been given by Kruuk (1964), and Tinbergen (1965); and Macdonald (1976) has illustrated this behaviour experimentally with tame red foxes storing mice. All these authors concluded that the function of this type of storing is to minimize the risk of wholesale losses of a large food source to other scavengers.

Brown hyaenas were observed to store, or to recover from storage, 65 food items (Table 3.11, Fig. 3.38). Similar caching behaviour has been recorded in brown hyaenas along the Namib Desert coast by Goss (1986), and in striped hyaenas in the Serengeti by Kruuk (1976). Of these 65 instances the subsequent behaviour was observed 12 times, excluding the first case related above involving the ostrich eggs. Four times the hyaena came back within 6 h, and four times within 24 h, to recover the food. Once, 48 h, and once, 72 h later, the food was gone and fresh brown hyaena tracks were found in the vicinity, although it was unknown whether the same hyaena that had stored had also retrieved the food. On another occasion, 72 h later, the food, in this case the hindquarters of a steenbok, had been taken by a leopard. The twelfth case was the longest period of food storing and involved a complete black-backed jackal carcass, which was stored under a small shepherd's tree. Two days later the carcass had been moved 200 m away by a brown hyaena and dropped in a tall grass clump. Most of the hair had fallen out, the carcass was crawling with maggots, and it stank. Four days later a brown hyaena carried this carcass to her den.

Seven observations were made of brown hyaenas coming to a bush, or a clump of grass, pulling out an old piece of hide, chewing on it for up to an hour and then putting it back. I never saw an animal do this more than

Table 3.11   Food items stored and type of site used by brown hyaenas.

Item Stored	Site			Total
	Bush tall	Clump of grass	Hole	
Herbivore leg	10	8	0	18
Ostrich egg	8	10	0	18
Piece of hide	2	5	0	7
Other meat or bone	15	5	2	22
Total	35 (54%)	28 (43%)	2 (3%)	65

(a)   (b)

Figure 3.38 Food storing by brown hyaenas. (a) The leg of a gemsbok concealed in a clump of grass. (b) The remains of a steenbok under a bush.

once at the same place, although it seems that they have a supply of this food source which they can draw on when other food is scarce. A comparable observation was made by Macdonald (1976), when a red fox came back on six occasions to chew on an apple core stored in a clump of grass.

3.7.2.2 SPOTTED HYAENA

Food storing in the spotted hyaena is a rare phenomenon (Table 3.12). The most common site used was a drinking trough at a windmill. Kruuk (1972) recorded water as the only storing site for spotted hyaenas in East Africa. It is interesting that this behaviour is also found in an area where there are no naturally occurring permanent waterholes. The other caching sites chosen are similar to those used by brown and striped hyaenas. The use of these sites and the associated behaviour of the spotted hyaenas point to the behaviour being rudimentary in this species, as is illustrated in the following examples:

18 February 1984. A hyaena carries the remains of a springbok lamb he has been eating to a fallen tree some 200 m away. He places the remains on the tree then lies down 50 m away. After an hour he returns to the remains and eats them.

4 June 1982. A hyaena chews briefly on the dried-out carcass of a hartebeest. She picks it up, and places it in amongst the branches of a fallen tree. She lies down close by for 12 min and then moves off.

Table 3.12   Food items stored and type of site used by spotted hyaenas.

Item stored	Site			Total
	Bush	Dead fallen down tree	Drinking trough	
Herbivore leg	1	0	7	8
Skin and bones	2	3	2	7
Springbok lamb	1	0	0	1
Total	4 (25%)	3 (19%)	9 (56%)	16

24 August 1980. An adult female picks up the head and skin of a gemsbok calf she and 11 others had been feeding on and carries them off, closely followed by a sub-adult. After 1.2 km she stops and places the remains under a bush, the second hyaena standing right behind her. After a minute the adult female picks up the remains again and carries them to another bush, leaving them under it. Immediately she moves off the sub-adult takes the food.

A spotted hyaena returned to the leg bone of a gemsbok approximately 12 h after it had been stored, and once, late in the afternoon, a hyaena retrieved from a drinking trough the skin of a wildebeest calf it had probably stored there during the night. Apart from these two cases, and the one where the hyaena returned after an hour to retrieve the springbok lamb remains, the length of time the food items were stored was not documented. Most of them were probably never recovered by the animals which originally stored them.

3.7.2.3 DISCUSSION
Brown hyaenas store food more often than do spotted hyaenas, being more like the similar striped hyaena in this regard. Macdonald (1976) and Andersson & Krebs (1978), hypothesized that the conditions promoting caching will be more rigorous in group-feeding than in solitary-feeding animals, as the chances of an individual regaining his own cache are smaller in a group-feeding species. This is illustrated in the third spotted hyaena example above. Furthermore, Macdonald (1976) suggested that caching is less likely to occur in species that are able to defend their food. As the group-feeding spotted hyaena is also able to defend its food far more efficiently than brown and striped hyaenas are, the behaviour of the Hyaenidae are compatible with these hypotheses.

Food storing by African carnivores is typically of short duration. The Hyaenidae usually return within 24 h to recover their food (Kruuk 1972, 1976, this study), as do black-backed jackals (Wyman 1967), and leopards (personal observations). This is in contrast to northern hemisphere carnivores, such as the red fox, and especially the Arctic fox and wolverine, where long-term storing occurs, particularly in winter (Ewer 1973). The African climate and numerous invertebrate scavengers do not allow meat to be preserved for long, and there are many other vertebrate scavengers which will steal cached food.

What then is the value of food storing to brown hyaenas in particular? Much of the food that they store, such as leg bones and old pieces of hide, is reasonably immune to the harsh climate and scavengers other than hyaenas. It is probably beneficial to remove the food item from its original source, such as a kill site, which would be expected to attract other scavengers, to a place further away under a bush, or in some tall grass, where the chances of it being discovered might be lessened. It usually costs little in time and energy to make a cache, so even a small advantage might make it worthwhile to do so. In the case of the ostrich nest recounted above this was not so; the hyaena spent nearly three hours hiding the eggs. Because they are so nutritious, long-lasting, and easy to hide, ostrich eggs appear to be particularly worthwhile items for brown hyaenas to store.

3.8 Summary

	Brown hyaena	Spotted hyaena
Activity period	Nocturnal	Nocturnal
Per cent of night active	80.2%	55.3%
Mean distance moved/ night	31.1 km	27.1 km
Day-time resting sites	Selective for shade and cover, particularly in summer	Less selective for shade and cover
Mean foraging group size	1	3 (Range 1–11) Variation depends on species hunted
Main senses used in foraging	Smell	Smell and sight
Foraging for fruits, insects and birds' eggs	Foraging adapted to finding them; can easily eat ostrich eggs	Poorly orientated towards them; eat ostrich eggs with difficulty
Hunting behaviour	Poorly developed, orientated towards small animals with a low success rate	Well developed, cursorial Mainly orientated towards large and medium-sized animals
Most common prey	Springhare, bat-eared fox, springbok lambs, korhaan	Gemsbok calves (63% success rate), gemsbok adults (14% success rate) and wildebeest (34% success rate)

| Feeding | Usually solitary | Often social, competition dependent on size of carcass and number feeding |
| Food storing | Common – usually in a grass clump or bush | Rare – usually in water |

Statistical tests

1. Comparison of the percentage of each hour between 18.00h–06.00h that brown hyaenas versus spotted hyaenas were active.
   Mann–Whitney U test: $U = 15.5$; $n_1 = 12$; $n_2 = 12$; $p <0.001$; one-tailed.
2. Comparison of the percentage of each hour between 17.00h–09.00h that southern Kalahari brown hyaenas versus Serengeti striped hyaenas were active.
   Mann–Whitney U test: $U = 68.6$; $n_1 = 17$; $n_2 = 17$; $p <0.05$; two-tailed.
3. Comparison of the mean distance travelled per night by brown hyaenas ($N = 42$) versus the mean distance travelled per night by spotted hyaenas ($N = 140$)
   Student's t test: $t = 1.437$; d.f. $= 180$; $p >0.05$; two-tailed.
4. Comparison of the distances that spotted hyaenas walked versus the distances they loped during foraging.
   Mann–Whitney U test: $U = 466.5$; $n_1 = 33$; $n_2 = 33$; $p > 0.05$; two-tailed.
5. Comparison of the mean detection distance from downwind to carrion ($N = 6$) by spotted hyaenas versus the mean detection distance from downwind to live prey ($N = 24$)
   Student's t test: $t = 5.616$; d.f. $= 28$; $p <0.001$; one-tailed.
6. Comparison of the mean detection distance by spotted hyaenas from downwind to live prey smelled through the air ($N = 24$) versus the mean detection distance scent trails to live prey were followed ($N = 11$).
   Student's t test: $t = 0.783$; d.f. $= 33$; $p> 0.05$; two-tailed.
7. Correlation between the mean distances spotted hyaenas chased gemsbok calves of varying ages in successful hunts and the age of the calves.
   Spearman rank correlation coefficient: $r_s = 1.000$; $p <0.05$; one-tailed; $N = 4$.
8. Comparison of the frequency with which one spotted hyaena was successful in gemsbok calf hunts versus the frequency with which two or more were successful.
   $\chi^2 = 0.53$; d.f. $= 1$; $p > 0.05$; $N = 39$.
9. Comparison of the frequency with which four or less spotted hyaenas were successful in gemsbok calf hunts versus the frequency with which five or more were successful.
   $\chi^2 = 3.46$; d.f. $= 1$; $p > 0.05$; $N = 39$.
10. Comparison of the mean distance gemsbok ran before facing hyaenas when the hyaenas made a determined effort to kill the gemsbok ($N = 15$) versus the mean distance gemsbok ran before facing hyaenas when the hyaenas did not make a determined effort ($N = 19$).
    Student's t test: $t = 2.067$; d.f. $= 32$; $p <0.05$; two-tailed.
11. Comparison of spotted hyaena hunting group size in successful hunts of adult gemsbok versus hunting group size in successful hunts of sub-adult gemsbok.
    Mann–Whitney U test: $U = 4$; $n_1 = 5$; $n_2 = 6$; $p = 0.026$; one-tailed.

12. Comparison of the mean distance wildebeest were chased by spotted hyaenas in successful hunts ($N = 16$) versus the mean distance they were chased in unsuccessful hunts ($N = 8$).
    Student's t test: $t = 2.396$; d.f. $= 22$; $p < 0.05$; two-tailed.

13. Comparison of the frequency with which groups of seven or more spotted hyaenas were successful in wildebeest hunts versus the frequency with which groups of six or less were successful.
    $\chi^2 = 13.31$; d.f. $= 1$; $p < 0.001$; $N = 51$.

14. Comparison of the frequency with which groups of 1–3 spotted hyaenas were successful in wildebeest hunts versus the frequency with which groups of 4–6 were successful.
    $\chi^2 = 0.13$; d.f. $= 1$; $p > 0.05$; $N = 51$.

15. Comparison of the frequency with which spotted hyaenas hunted springbok during January and February versus the frequency with which they did so between March and December.
    $\chi^2 = 33.8$; d.f. $= 1$; $p < 0.001$; $N = 41$.

16. Comparison of the frequencies with which spotted hyaenas from different clans were involved in successful hunts.
    $\chi^2 = 12.67$; d.f. $= 5$; $p < 0.05$; $N = 136$.

4 Social structure and spatial organization

In the last two chapters I discussed the feeding ecology and behaviou the hyaenas. Now I examine and compare their social structure and sp organization, and the influence of diet and food dispersion in moul them.

First I consider groups. Most individuals of both species live in f territories, with matrilineally related conspecifics and fewer, unrel immigrants. These groups were first documented in the spotted hyaen Kruuk (1972), who called them clans. Other individuals, mostly males not belong to a clan and are nomadic. The importance of these anin particularly in the brown hyaena's mating system (Mills 1982b), become apparent in Chapter 7. Clans are dynamic; cubs are born individuals immigrate; members are lost through death and emigrat What is the contribution of each of these four factors to clan dynamics?

In the second part of the chapter I consider the land tenure systen the hyaenas and their movement patterns within their territories (I 1982a, Mills & Mills 1982). This leads to important questions about fac affecting group size and territory size – and, ultimately, the siz population.

4.1 Brown hyaena clans

4.1.1 Clan structure and fluctuations

Numbers fluctuated markedly, both between and within brown hy clans (Fig. 4.1, Table 4.1). The largest number of adults and sub-adults

Table 4.1   The composition of five brown hyaena clans.

Group	Date	Adults		Sub-adults		Adult or sub-adult of unknown sex	Cubs	Total
		M	F	M	F			
Cubitje Quap	August 1972	1	1	0	0	0	2	4
Kaspersdraai	May 1973	0	1	0	0	0	3	4
Rooikop	December 1973	1	1	0	0	1	2	5
Seven Pans	May 1975	?	2	1	0	2	1	6
Botswana	May 1975	2	1	0	2	0	1	6

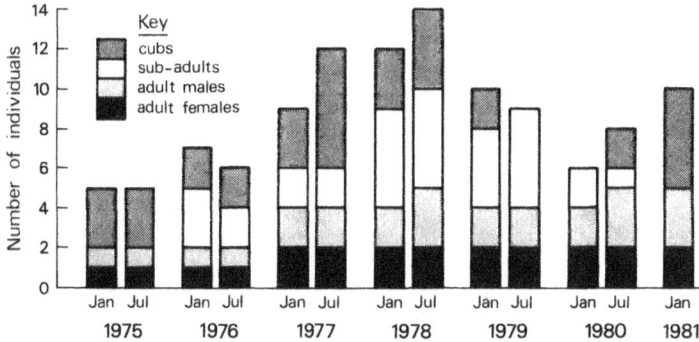

Figure 4.1 The structure of the Kwang brown hyaena clan at six-monthly intervals, January 1975–January 1981.

clan was 10, in the Kwang clan in July 1978 (Fig. 4.1), and the smallest observed clan comprised a single adult female and her three cubs (Table 4.1). The fluctuations in the Kwang clan were due mainly to births and the disappearance of sub-adults. Adult numbers remained relatively constant (Fig. 4.1).

Most clans had only one adult female, although in some there were two. Owens & Owens (1979a) recorded as many as five females in a brown hyaena clan in the central Kalahari. Overall, the sex ratio of 19 adults from six clans was 1.4:1 (males:females), which does not deviate significantly from a 1:1 ratio (1).

4.1.2   Recruitment

The potential rate of cub production by brown hyaenas was not realized, mainly because females rarely produced cubs within the shortest possible time periods (see for example Fig. 4.2). The observed periods between six litters born to three females fluctuated from 12 to 41 months, with the mean being 20.7 ± 4.9 months. The only known age of first parturition was 35 months. There was little evidence for seasonality in births in the small sample (Fig. 4.3) (2).

Fluctuations in litter size were unimportant determinants of group size. The mean size of nine litters of cubs under six months old when first found was 2.7 ± 0.3, with the mode 3 and the range 1–4. In captivity the mean size of 35 litters born worldwide was 2.1 (Shoemaker 1978, 1983).

4.1.3   Emigration

Proximately, emigration was the most important determinant of brown hyaena group size. Both sexes, but particularly males, emigrated from their natal clans. Of the 15 cubs born into two clans between 1974 and 1978

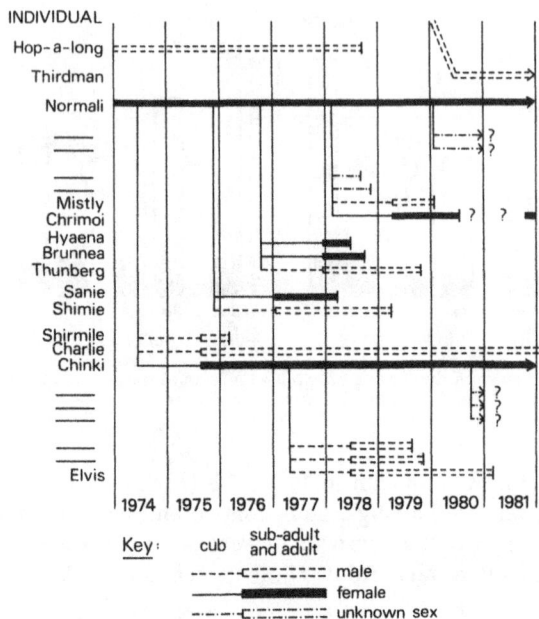

Figure 4.2 The reproductive history of the Kwang brown hyaena clan, 1974–1981. Birth of a litter is shown by a vertical line descending from the female's line. The ending of a horizontal line by one short vertical line denotes either death or emigration. An oblique line denotes immigration. Named individuals are mentioned in the text.

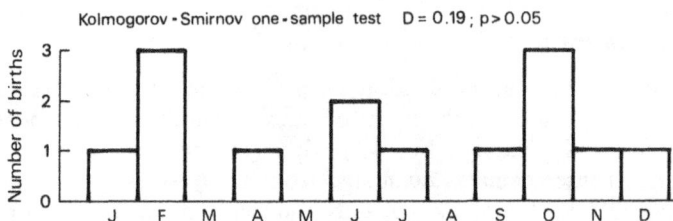

Figure 4.3 Number of litters of brown hyaenas born each month of the year throughout the study.

and raised to sub-adulthood, all except three females eventually disappeared, two of which were still breeding in their natal territory when they were over ten years old. The third female (Chrimoi – see Fig. 4.2) disappeared from her natal clan between July 1980 and late 1981, after which she was seen regularly again in her natal territory. Although the age at disappearance fluctuated between 20 and 78 months, and was the same for males and females (3), eight of the 12 animals disappeared as sub-adults, i.e. between 20 and 30 months.

I do not know what became of the three female emigrators, but of five males which left the Kwang clan, two apparently became nomadic and three eventually joined other clans. The two that became nomadic were Charlie, the oldest animal to leave a clan at 78 months, and Elvis, the second oldest at 45 months (see Fig. 4.2). Charlie was seen twice in the Kwang territory, 17 and 30 months after leaving, and Elvis was seen once in the Kwang territory, 19 months after leaving. The three that joined other clans were Shimi, who disappeared at 41 months and joined the neighbouring Rooikop territory (Fig. 4.13) three years later, and Thunberg and Mistly, which joined separate clans along the Nossob river bed.

It appears, therefore, that most, if not all, brown hyaena males eventually leave their natal clans, although some may only do so long after reaching adulthood. Some brown hyaena females probably remain for life in their natal clans, but others leave, after which it is unknown what happens to them.

4.1.4   Immigration

Immigration was unimportant in regulating group size. Only males were known to immigrate. Besides the three males mentioned above, the male, Thirdman, first appeared in the Kwang territory in June 1980 (Fig. 4.2) as an age-class 3 (see section 1.3) animal, and stayed in the area until at least April 1984, when I left the Kalahari. Another male, caught and marked as a sub-adult in December 1974, was seen five times in an area 45 km northwest of the capture point between May 1981 and August 1982, and he too most probably had joined the local clan.

4.1.5   Mortality

Cub mortality was low. Of 19 cubs first seen under six months old, only two (11%) died before they became sub-adults. I was unable to measure accurately sub-adult (age-class 2) mortality, but as there was a sharp drop from the percentage of age-class 2 animals in the population to the percentage of age-class 3 animals (Fig. 4.10a), and as age-class 2 animals died in slightly higher proportion to their incidence in the population (Fig. 4.10b), it appears to have been comparatively high. Most adults (age-classes 3–5) died in old age (Fig. 4.10b). Moreover, the large number of age-class 4 (6–16 years) animals in the population (Fig. 4.10a) suggests that it was demographically healthy.

Table 4.2 documents the known cases of brown hyaena mortality. Those killed by man were from farms adjoining the two National Parks in Namibia and Botswana, or were carcasses found in gin-traps, or poisoned along the border fences (E. le Riche personal communication). Although mortality due to man was probably over-represented, because rangers' patrols were mainly along boundary fences, it nonetheless appears to be an

Table 4.2   Causes of mortality of brown hyaenas from different age-classes. Unknowns are hyaenas that were found too long after death for the cause to be determined, or known cubs that disappeared. The proportion in which age-class 1 animals appear is not representative.

Cause	Age-class						Total
	1	2	3	4	5	Unknown	
Man	2	6	0	1	0	5	14 (37%)
Violent	1	3	1	1	4	3	13 (34%)
Unknown but natural	5	1	1	0	4	0	11 (29%)
Total	8 (21%)	10 (26%)	2 (5%)	2 (5%)	8 (21%)	8 (21%)	38

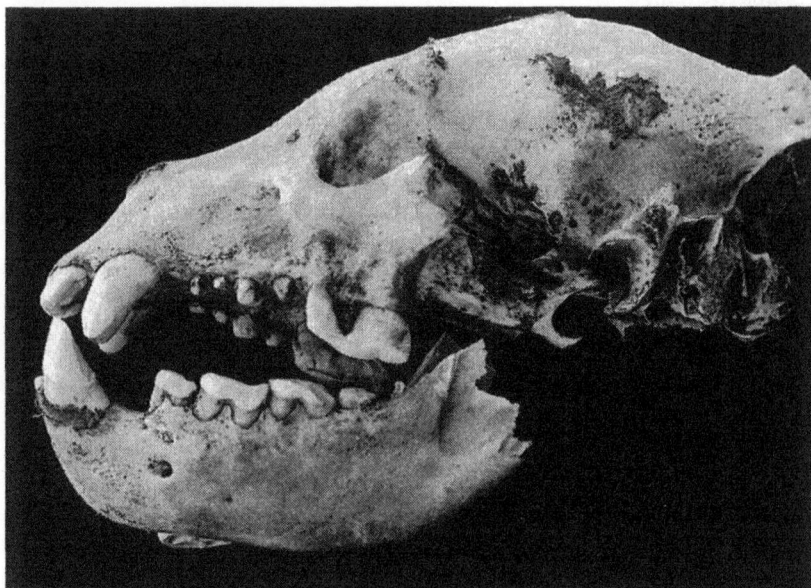

Figure 4.4   The skull of an age-class 5 brown hyaena. Note that the premolars in the upper jaw particularly have become so worn as to be non-functional.

important mortality factor. The majority of brown hyaenas killed in this way were age-class 2 animals – those most likely to emigrate.

Violent mortality in the brown hyaena seems to be the most common form of natural mortality. Most of the wounds inflicted were along the back and around the hind region. Such injuries were most probably caused by other species of large carnivore, rather than by other brown hyaenas, although in the 13 instances where a brown hyaena died after being attacked, lions could only be identified as being responsible on three occasions, and spotted hyaenas once. Two brown hyaenas died through severe neck injuries. These may have been caused through intra-specific fights, as fighting between brown hyaenas, although usually inhibited, tends to be directed at the neck (section 5.1.2).

Wear eventually causes the bone-crushing premolars of old brown hyaenas to become non-functional, which must put them at a severe disadvantage, causing starvation, or starvation-related mortality. For example, an old male was observed to spend an unsuccessful 30 min attempting to crush the leg bone of a springbok, a feat which a young animal would have accomplished within five minutes. Three months later this individual died. His skull is shown in Figure 4.4.

Little is known about the roles which disease and parasites play in brown hyaena mortality in the wild. Rabies, which may have been important for spotted hyaenas (section 4.2.5), was never suspected.

4.2   Spotted hyaena clans

4.2.1   Clan structure and fluctuations

The composition of several spotted hyaena clans is shown in Figure 4.5. Less detailed observations on two other clans (Leijersdraai and Kanna-gauss) are not illustrated, but suggest that they comprised about ten individuals each. With the exceptions of the Kaspersdraai, Seven Pans and Urikaruus clans, therefore, the other four clans, including the Kousaunt clan for 6½ years, contained eight to 13 adults and sub-adults, and usually there were between nine and 11. There tended to be less fluctuation in the size of spotted hyaena clans than in brown hyaena clans.

Typically, there was only one adult male in a clan, but between November 1980 and August 1983 up to four adult males were present in the Kousant clan (Fig. 4.5a). However, two were young adults whose departure from the natal clan was slightly delayed, and the third an immigrant (Nicholas) (Fig. 4.6) which was only peripherally associated with the clan. Only the fourth male (Hans), also an immigrant (Fig. 4.6), was fully integrated into the clan for an extended period (see section 7.3.2.1). No adult male was observed to associate with the Seven Pans animals over a 15-month period, although four adult males were seen in the area (section 4.3.2).

The number of resident adult females was also relatively constant. There were six in the Kousaunt clan for the latter half of 1982 and the first half of 1983. Then the clan underwent a fission (Fig. 4.6 and section 7.3.3.3), which was apparently more related to the number of females in the clan than the total number of animals (Fig. 4.7). In April 1986 there were again six adult females in the clan and two of these departed (M. Knight personal

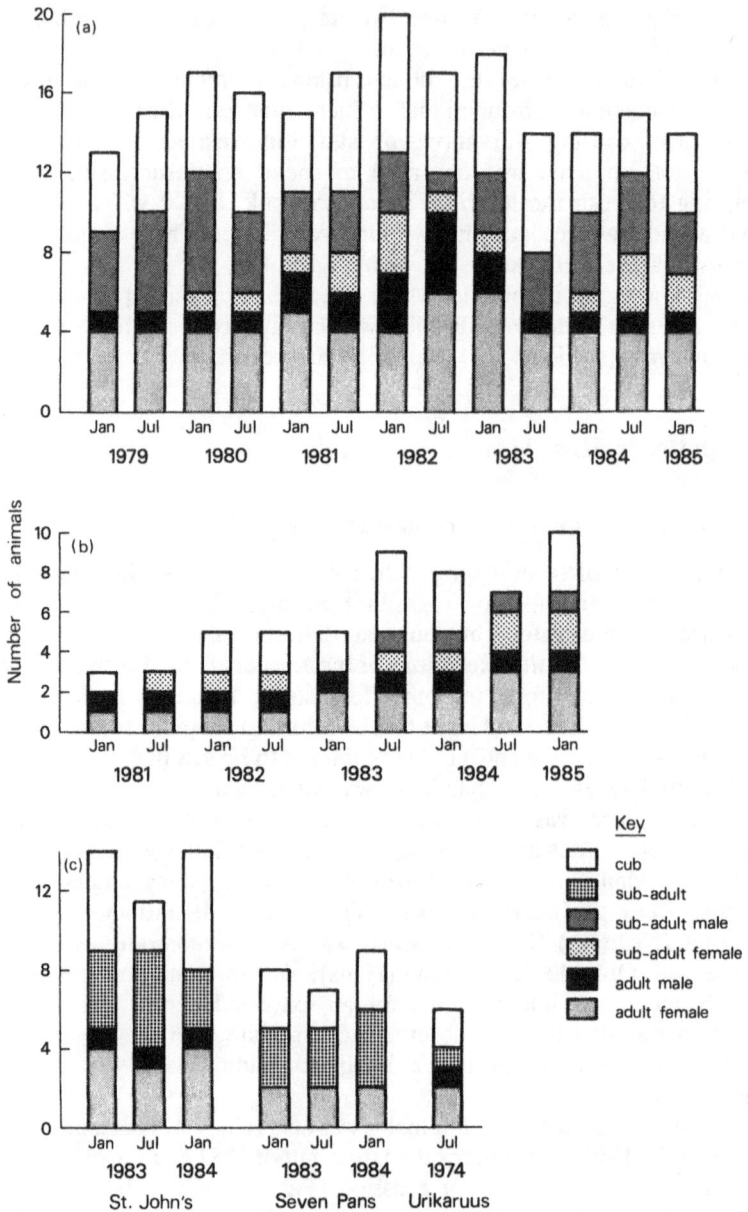

Figure 4.5 The structure of various spotted hyaena clans at six-monthly intervals. (a) The Kousaunt clan. (b) The Kaspersdraai clan. (c) The St John's, Seven Pans and Urikaruus clans.

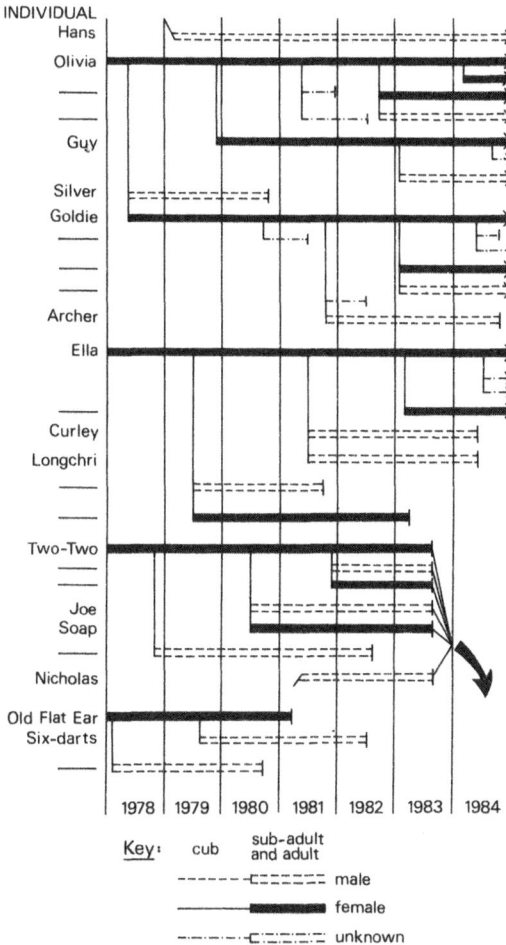

Figure 4.6 The reproductive history of the spotted hyaena Kousaunt clan, 1978–1984, showing the clan fission in August 1983 by means of the curved arrow. For interpretation of symbols see Figure 4.2.

communication). Mostly, therefore, there were four breeding adult females in this clan, which was also the maximum long-term number of breeding females in all clans (Fig. 4.5).

In contrast to the brown hyaena where the sex ratio in clans was not significantly different from unity, the sex ratio in spotted hyaena clans was 0.4:1 (males:females) (4).

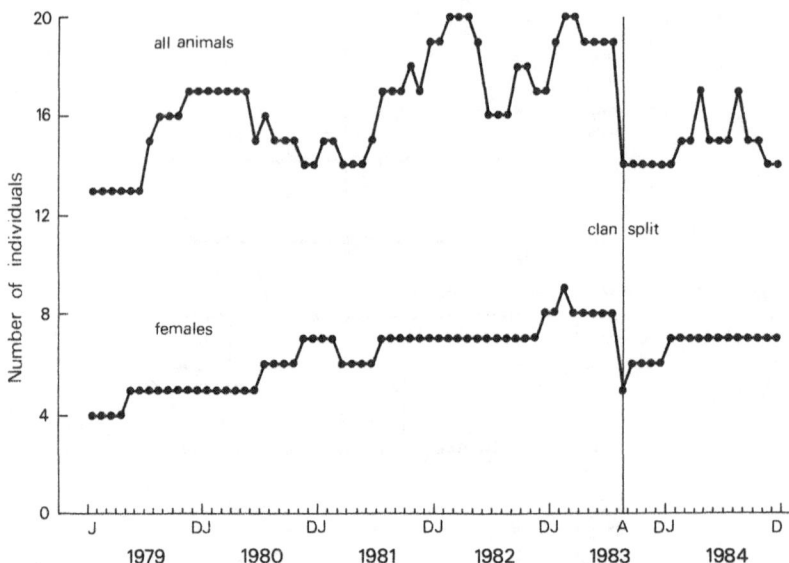

Figure 4.7 The number of all individuals, and females only, in the Kousaunt spotted hyaena clan each month from January 1979–December 1984.

4.2.2 Recruitment

As with brown hyaenas, the potential rate of cub production in spotted hyaenas was not realized. The mean interval between 27 litters was 16.7 ± 0.8 months, which is similar to the 17.6 months recorded by Kruuk (1972), but somewhat longer than the 14.6 months recorded by Frank (1986a), in the Mara National Reserve. However, if cases where cubs died are excluded, the mean was 18.8 ± 0.5 and the minimum 15 months. In all cases where a female lost a litter she produced the next three to five months later (see also section 7.3.3.1).

The mean age at which five females gave birth for the first time was 37.0 ± 2.4 months (range 30–45 months). A female which gave birth at 38 months was observed in courtship behaviour when she was 33 months old, but apparently did not conceive. Three females had not given birth to young at 37, 38 and 45 months. Frank (1986a) recorded a mean age of first parturition of 36.6 months.

Spotted hyaena litter sizes are small. The mean of 20 litters with cubs under one month old when first observed, was 1.7 ± 0.1, with 14 of the litters containing two cubs and six containing one cub. Of nine litters comprising two cubs, eight consisted of a male and a female, and one consisted of two males, which is similar to Frank's (1986b) data. Of nine pregnant females collected in the Kruger National Park, eight contained two foetuses and one contained a single foetus, giving a mean of 1.9

Kolmogorov - Smirnov one - sample test   D = 0.1 ;   p > 0.05

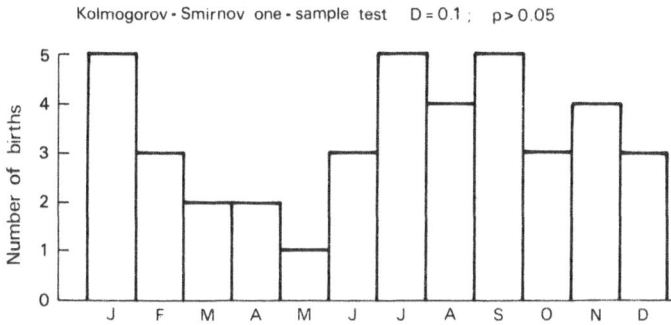

Figure 4.8 Number of litters of spotted hyaenas born each month of the year throughout the study.

(Lindeque 1981), and data from captivity give a mean litter size of 1.97 (Schneider 1926, Pournelle 1965, Crandall 1964, Golding 1969).

Although not marked (5), there was a slight trough in births between March and June (Fig. 4.8). Similar seasonality of births has been recorded from other spotted hyaena populations (Kruuk 1972, Lindeque & Skinner 1982a, Frank 1986a).

4.2.3 Emigration

Emigration was also the most important determinant of group size in the spotted hyaena, although there were differences in emigration patterns between males and females. All of 15 male cubs followed through to sub-adulthood disappeared from their natal clans at an average age of 35.2 ± 1.2 months (range 30–45 months). This is about six months older than the average age recorded by Frank (1986a). Five disappeared during the last year of the study, so the chances of contacting them again were limited. Of the remaining ten, six, all from the Kousaunt clan, were subsequently seen at least once:

1. Two briefly rejoined their natal clan ten and 11 months after disappearing and were fully accepted by the other members of the clan, although neither stayed.

2. A third was seen once in the company of two adult females from the St John's clan, approximately 125 km SSE from Kousaunt (Fig. 1.1), ten months after disappearing, but was not seen again.

3. A fourth hyaena was found two months later in his natal territory at Kwang (Fig. 1.1), with a gin-trap on his front foot (section 4.2.5). The nearest place he could have picked up the trap was 55 km to the west along the Namibian/KGNP boundary.

4. The fifth hyaena (Six Darts) departed in June 1982, and moved extensively for at least 15 months thereafter, after which he was not seen again (Fig. 4.13).

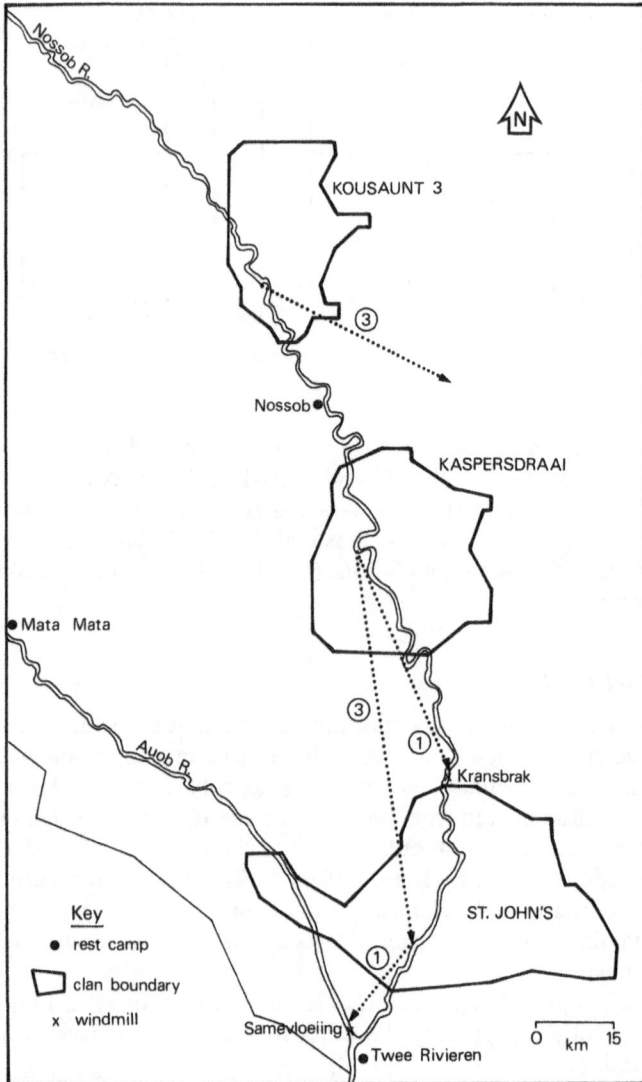

Figure 4.9 Four cases of female spotted hyaenas changing clans. The numbers refer to the number of females involved in each case and the known boundaries of three clans in 1983 are shown.

5. The sixth hyaena (Silver) disappeared in October 1980 at 30 months of age. Twenty-two months later he had joined the St John's clan, and remained a member until at least February 1986 (M. Knight personal communication).

Females were more likely to be integrated into their natal clan than males, as was also found in other spotted hyaena studies (Frank 1986a, b, Henschel & Skinner 1987), and with brown hyaenas (section 4.1.3). Of four females in the Kousaunt clan which reached adulthood during the study, two stayed with the clan and bred (Fig. 4.6). Subsequently, another two females born into the clan have bred there as well (M. Knight personal communication). In the Kaspersdraai clan both of the females that reached adulthood during the study remained with their natal clan and bred, and a female born in the St John's clan also remained and bred.

Some females did leave their natal clan:
1. A female (Two-Two) and her two daughters and two sons, together with an immigrant male, split off from the Kousaunt clan (Fig. 4.6 and section 7.3.3.3), and established a new clan to the southeast.
2. Three females from the Kaspersdraai clan established themselves in the St John's area (Fig. 4.9), although I do not know if they had joined a clan, or established a new one, and a fourth female from the Kaspersdraai clan joined the Kransbrak clan (Fig. 4.9). Throughout this time spotted hyaenas were still seen regularly in the Kaspersdraai area.
3. Three and a half years after moving to the St John's area, one of the three females that originated from Kaspersdraai, moved again to join the neighbouring Samevloeiing clan (Fig. 4.9). She was not the only adult female member of this clan, but I do not know the origins of, nor the relationships between the Samevloeiing animals. She remained in this area until her death 11 months later.

These observations are reminiscent of Kruuk's (1972) in the Serengeti, where spotted hyaenas of either sex and from different areas came together to form clans, although the Serengeti clans were merely temporary associations.

4.2.4 Immigration

During the six-year study period the only animal to join the Kousaunt clan was an age-class 3 male (Nicholas, see Fig. 4.6), although he never integrated completely into the clan (section 7.3.2). The male, Hans, was also probably an immigrant male, which had joined the clan shortly before the commencement of intensive observations (section 7.3.2).

Three immigrant males associated with the Kaspersdraai clan between 1980 and the end of 1985. One (Cyclops), was present from 1980 to October 1982, the second only for September and October 1982, and the third remained from November 1982 until at least October 1985 (M. Knight personal communication). As has been mentioned, a male (Silver) from the Kousaunt clan transferred to the St John's clan and remained for at least three and a half years as a fully integrated member of the clan. Although male immigration was unimportant in regulating group size, immigrant males are important in the mating system of the spotted hyaena (Ch. 7).

Female immigration was even less common, but may have taken place when the Kaspersdraai animals moved to the St John's area, the St John's animal moved to Samevloeiing, and the Kaspersdraai animal moved to Kransbrak (section 4.2.3). In the Namib Desert, Tilson & Henschel (1986) documented a case of an adult female, together with a sub-adult male, joining a clan.

4.2.5 Mortality

As with the brown hyaena, spotted hyaena cub mortality was low. However, adult mortality patterns were different from the brown hyaena's, with animals dying more frequently in their prime.

During the study, 27 (75%) of the 36 spotted hyaena cubs first seen under one month old reached sub-adulthood. Sub-adult mortality was also low while they were with their natal clans; only one of 23 (4%) animals observed from 15 through to 30 months of age died.

Figure 4.10a shows that the age structures of adult animals in the southern Kalahari spotted hyaena population differs from other hyaena populations. There were relatively more individuals in age-classes 2 and 3, and fewer in age-class 4 in this population. These data are substantiated in Figure 4.10b, which shows that proportionally more spotted hyaenas from age-classes 4 and 5 died in the southern Kalahari than in the other populations (6). The southern Kalahari spotted hyaena population, there-fore, appears to be a younger population than the others. It seems unlikely that this is due to methodological problems in ageing hyaenas, because of different rates of tooth wear between Kalahari and East African ones. If anything Kalahari hyaenas' teeth should wear down quicker than East African animals', because of the abrasive action of the sand in the Kalahari (Smuts et al. 1978). Therefore, Kalahari hyaenas may be younger than their East African counterparts at any particular stage of tooth wear. If this is so it means that the Kalahari population is even younger than the East African ones.

Can mortality factors explain this aberrant population structure in Kalahari spotted hyaenas? It was often impossible to establish the cause of death (Table 4.3). As with brown hyaenas, several were killed by farmers along park boundaries – four were shot at one time, one was poisoned, and another (referred to in section 4.2.3) had to be put down after its foot had been caught in a gin-trap which cut through the bone. (The injuries of two other spotted hyaenas which survived were also probably due to gin-traps. One of them (the male Cyclops in Fig. 4.13) became almost fully mobile again after two years. The second appeared to be coping adequately with his handicap on the few occasions he was seen.)

Of the violent mortality cases in Table 4.3, two cubs were killed by lions (see Fig. 2.27), and spoor revealed that an age-class 3 animal had been killed by other spotted hyaenas (D. de Villiers personal communication).

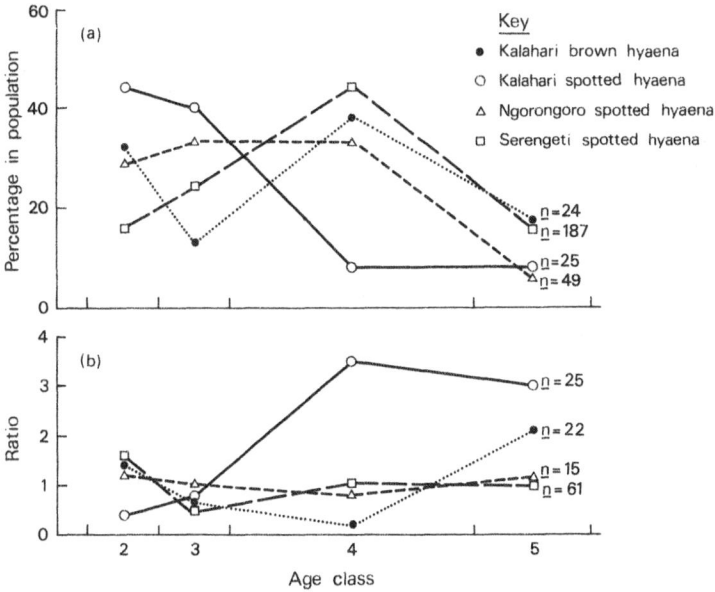

Figure 4.10 (a) Age-class distribution based on tooth wear of randomly caught hyaenas from four populations. (b) The ratio of the percentage of hyaenas found dead to the percentage in the population for different age-classes from these four populations. The data from the Serengeti and Ngorongoro are from Kruuk (1972). The age-class intervals are proportional to the time that animals spend in each age-class.

Table 4.3   Probable causes of mortality of spotted hyaenas from different age-classes. Unknowns are the same as in Table 4.2.

Probable cause	1	2	3	4	5	?	Total
Rabies	3	0	1	3	0	5	12 (24%)
Man	0	2	2	0	1	1	6 (12%)
Violent	2	0	1	1	0	2	6 (12%)
Paralysis	0	1	2	0	0	0	3 (6%)
Unknown disease or starvation	0	0	0	0	0	1	1 (2%)
Unknown	10	1	2	3	5	1	22 (44%)
Total	15 (30%)	4 (8%)	8 (16%)	7 (14%)	6 (12%)	10 (20%)	50

I was not able to identify the attackers in the other cases. Kruuk (1972) recorded violent mortality, mainly by lions, but also by other spotted hyaenas, as the prime mortality factor for adult spotted hyaenas in East Africa.

The largest suspected mortality factor, however, was rabies. Of the 28 cases where a likely cause of death could be found, 43% were attributed to rabies (Table 4.3), although only one was confirmed. Two suspected outbreaks of rabies were observed in the southern Kalahari between 1972 and 1987. The first was in September 1974 when I made the following observation:

22 September 1974. 17.00 h. I arrive at the Urikaruus spotted hyaena den to find an adult female and a sub-adult lying there with blood around their faces. Suddenly the sub-adult jumps up, and attacks the adult female, biting at her hindquarters. The female moves away, but when the sub-adult starts lowing (see Table 5.6), she runs over to it and they clash again. A second adult female appears. The two females greet and one of them whoops several times. Then the sub-adult runs, while fast whooping (Table 5.6), at a small cub which is emerging from the den. The cub quickly retreats and the sub-adult spins around and attacks the adult females. It bites at their hindquarters, and the females take evasive action, but they make little effort to counter-attack (Fig. 4.11). The sub-adult then bites at their necks and again they take evasive action. Then it attacks their mouths, at which the females counter-attack, one of them grabbing the sub-adult by the side of the mouth and shaking it vigorously for several seconds. An adult male is seen approaching the den. The sub-adult rushes towards him, but the male runs away. The sub-adult continues running until it comes to Urikaruus windmill, 2 km

Figure 4.11 Two adult female spotted hyaenas (left) defend themselves from the attack of a sub-adult (right) suspected of having rabies.

away. It stands in the drinking trough and attempts to drink. However, it is only able to scoop a little water into its mouth. The animal has a dazed look about it and tends to stumble.

This was the last time the sub-adult was seen, in spite of several nights' observations at this den over the next month. The remainder of the clan, the two adult females, their two cubs and the adult male (Fig. 4.5c), had all disappeared two months later. It seems likely that this small clan was eradicated by rabies, as transmission is generally through being bitten by an infected animal (Kaplan 1977).

The second suspected rabies outbreak took place in early 1986 (M. Knight personal communication). Six animals died of suspected rabies between January and May in the northern Nossob region. One was reported to have attacked cars, two had apparently fought with each other and one had attacked a fibreglass drinking trough at a windmill. The brain of this last-mentioned animal was collected and was found to be positive for rabies. Two of the animals were from the Kousaunt clan, the immigrant male, Hans, and the dominant female, Olivia (see Ch. 7). Follow-up observations revealed that these were the only adults from this clan which succumbed to rabies during the outbreak (M. Knight personal communication).

Chapman (1978) recorded how rabies decimated a wolf pack in Alaska. He pointed out that since members of a wolf pack are socially close, most, if not all, of the members will be exposed to rabies, should one member of the pack contract the disease. Furthermore, as packs are discrete social units, the chances of rabies spreading from one pack to another, especially in a low-density area, must be lower than within packs. The same should hold true for southern Kalahari spotted hyaenas and seemed to occur in the first outbreak recorded above, but not in the second one. However, as Macdonald & Bacon (1982) have shown with red foxes, the flexible behaviour of carnivores can cause differences in frequencies of encounters between individuals, and hence in potentially different contact rates for rabies. Moreover, the behaviour of animals which have contracted rabies may vary (Barnard 1979); sometimes they may seek a familiar area, at other times they may become abnormally wide-ranging (V. de Vos personal communication).

There is, therefore, some evidence, albeit sketchy, that rabies may sometimes cause reductions in spotted hyaena numbers in the southern Kalahari. The clan that disappeared from Urikaruus was not replaced for at least seven years, so endemic rabies may contribute to the low spotted hyaena density in the southern Kalahari. If older (age-class 4) animals are more vulnerable to rabies, perhaps because of their central position in the social life of the clan, and consequently more frequent interactions with other group members, it may explain the unusual age structure and mortality ratios of the southern Kalahari spotted hyaena population in comparison with others (Fig. 4.10). The fact that rabies has not been found in the brown hyaena can possibly be explained by the more solitary existence of this species, and the reduced probability of a rabid individual contacting another conspecific. Alternatively, it could be due to differential susceptibility to the strain of virus.

4.3 Nomadic males

Observations of some southern Kalahari hyaenas of both species suggested a nomadic existence; i.e. they travelled through several territories with apparent disregard to territory borders, and did not have any lasting relationships with other conspecifics. In both species these nomads were mainly males of all ages, but were more common in the brown hyaena population than the spotted hyaena population.

Amongst carnivores, nomads characteristically have low reproductive success (Mech 1970, Schaller 1972). However, as I will show in Chapter 7, in hyaenas nomadic males may be reproductively active, and in the case of the brown hyaena are of particular importance in the mating system. In this section the existence of nomadic males in the social systems of both species is documented.

4.3.1   Brown hyaenas

Although circumstantial, evidence for some male brown hyaenas being nomadic comes from several sources:
1. Figure 4.12 shows the number of times that I resighted all recognizable adult brown hyaenas in known clans' territories. This shows a discontinuous distribution with many individuals never being resighted, or

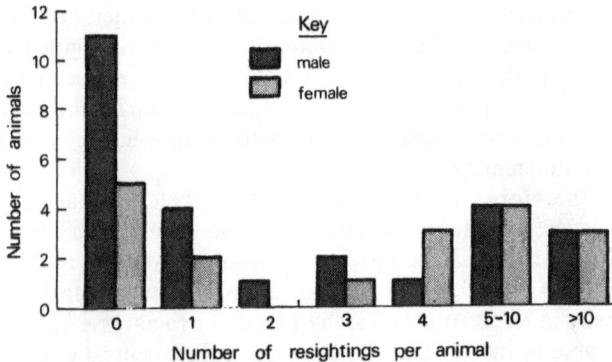

Figure 4.12 The number of times 44 individually recognizable adult brown hyaenas were resighted in the territories of known clans between 1975 and 1983, provided that they were seen for the first time before the end of 1980.

resighted only once, a few being resighted two or three times and more again being resighted four or more times. Radio tracking and trapping data, as well as observations at dens, showed that at least 14 of the 18 individuals that were resighted four times or more remained in the territories in question for a minimum of one year, or were known to have been born there. All these animals, therefore, were assumed to be members of the resident clan. None of the animals sighted less than four times fitted either of these criteria, therefore it seems that most, if not all of them, were not members of a clan.

2. Of animals resighted less than four times males outnumbered females by 2.3:1 (7), whereas the sex ratio of animals resighted four times or more was close to unity (0.8:1) (8), as was the sex ratio of adults from known clans (Section 4.1.1).

3. The behaviour of animals after they were caught and marked showed that animals of age-class 2 were likely to disappear from the area in which they were marked, but that few age-class 3 or older animals did, the only ones doing so being males (Table 4.4).

It is possible that these transients were animals from other clans, but the evidence does not support this hypothesis. Although there is overlap between the territories of neighbouring clans (Fig. 4.14), and both males and females occasionally made short expeditions out of their territories, only two such expeditions were observed in over 900 km of following males. Furthermore, these trips covered comparatively small distances (17 and 6.5 km), and appeared to be normal foraging expeditions. In addition, in five years I never saw any of the four marked resident males from territories neighbouring the Kwang clan outside their territories, nor in association with any of the Kwang animals.

A more likely hypothesis is that these males were nomadic. Of the 18 males that were identifiable and which were not known to be group

Table 4.4   The subsequent behaviour of 36 brown hyaenas after they were captured and marked.

Subsequent behaviour	Number of animals	Age-class 2		Age-class ≥ 3	
		Male	Female	Male	Female
Remained in the area in which it was marked for at least one year	23 (64%)	6	5	6	6
Disappeared within three months after being marked	13 (36%)	7	3	3	0
Total	36 (100%)	13	8	9	6

members, 11 were never resighted and seven were resighted at least once within a month of the original sighting in more or less the same area. After this, four of the latter seven were never seen again, and three were resighted once each, 14, 18, and 27 months later, again in the same area as the original sightings. In addition, there were the observations of the two

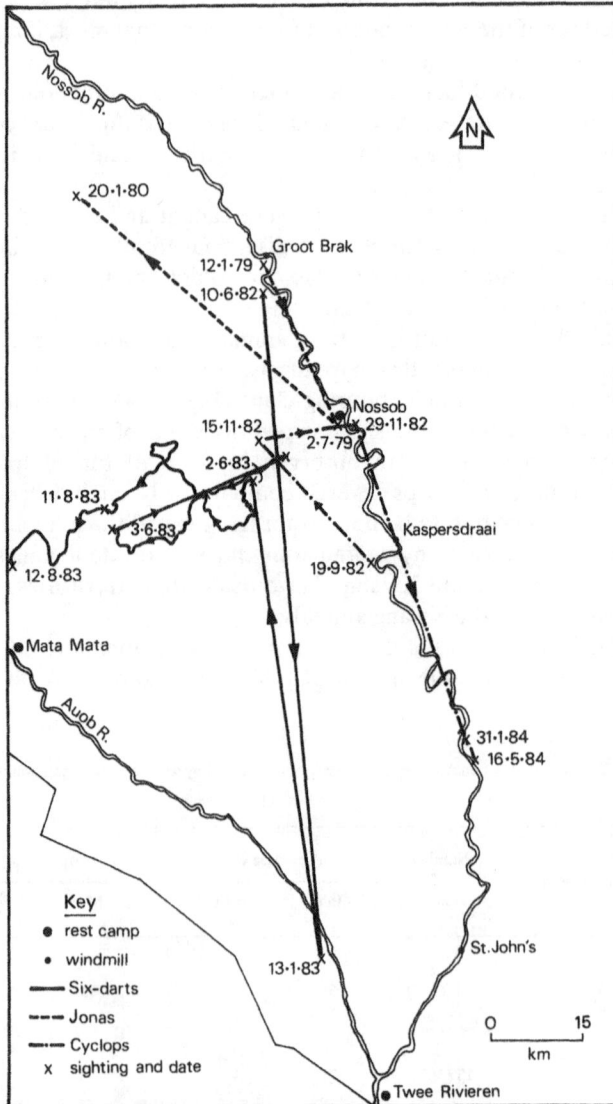

Figure 4.13 The movements of three nomadic male spotted hyaenas. Straight lines connect successive sightings, crooked lines show actual movements.

males that reappeared in their natal clan territories briefly after departure (section 4.1.3). Finally, a male was seen once in the Kaspersdraai territory and five weeks later in the Kwang territory, 39 km away, and never again.

The data from captured animals suggest that nomadic males account for 8% of the sub-adult and adult segment of the population, and 33% of the adult male segment of the population (Table 4.4). Adult females were less inclined to be nomadic, although there was some evidence for a female (Chrimoi) being temporarily nomadic (section 4.1.3).

4.3.2 Spotted hyaenas

Several observations were made supporting the hypothesis that some male spotted hyaenas were also nomadic.

1. When I commenced observations on the Kousaunt clan in January 1979, there was an age-class 5 male (Jonas) associating with the clan, but he was rejected by the other clan members (see section 7.3.2.1). He quickly disappeared from the clan. Five months later he was seen twice around Nossob camp over a 30 day period, 38 km from his last sighting. Six months after this he was seen once near Dankbaar, 63 km from Nossob (Fig. 4.13). During this period he apparently did not belong to any clan and lived a solitary, nomadic existence.

2. The movements of the males, Six Darts and Cyclops (after he departed from the Kaspersdraai and before he possibly joined the Kransbrak clan), were also of a solitary, nomadic nature (Fig. 4.13).

3. As mentioned earlier, four adult males (Cyclops, Six Darts and two others), were seen once each in the Seven Pans territory over a 15-month period.

4. Of 13 identifiable individuals which were sexed and which were not known to be attached to a clan, 12 were males. Of these, at least three were age-class 2 animals, and at least one of each of the others belonged to the other three adult age-classes. Two others, which could not be aged accurately, were obviously old (age-class 4 or 5) animals.

Spotted hyaena nomadic males do not appear to form such a large proportion of the population as in the brown hyaena. Similar observations of nomadic males were made in the Masai Mara National Reserve by Frank (1986a), who called them transients.

4.4 Land tenure system

4.4.1 Territories

The land tenure systems of Kalahari hyaenas meet the criteria of most authors for a territorial system (see for example Brown & Orians 1970, Wilson 1975, Davies 1978):

- The members of a clan inhabit essentially the same area.
- Each clan's area is largely exclusive (Figs 4.14, 4.15). (Even in areas of overlap one of the clans would utilize the area more than the other and they would avoid each other temporarily.)
- The areas are scent marked.
- Hyaenas from neighbouring clans behave aggressively towards each other (sections 7.2.4, 7.3.4).

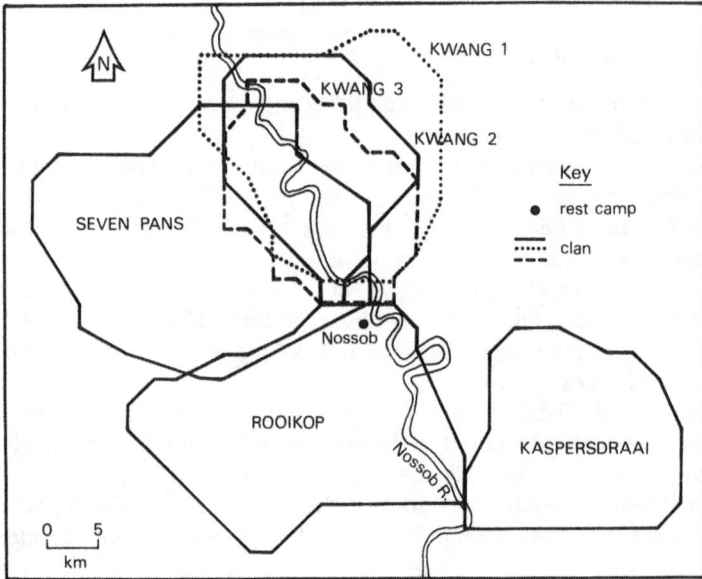

Figure 4.14 Restricted polygons enclosing the observed movements of brown hyaenas from four clans. The Kwang clan is shown for three time periods: Kwang 1 1975; Kwang 2 1976; Kwang 3 1977–1978.

Table 4.5   Brown hyaena and spotted hyaena territory sizes (km^2) as measured by the restricted polygon method (see Appendix D).

Brown hyaena		Spotted hyaena	
Kwang 1*	297	Kousaunt 1**	989
Kwang 2	228	Kousaunt 2	1046
Kwang 3	215	Kousaunt 3	553
Kaspersdraai	267	Kaspersdraai	814
Rooikop	381	St John's	1394
Seven Pans	461	Seven Pans	1776
Mean ±- SE	308 ± 39		1095 ± 177

* see Figure 4.14
** see Figure 4.15

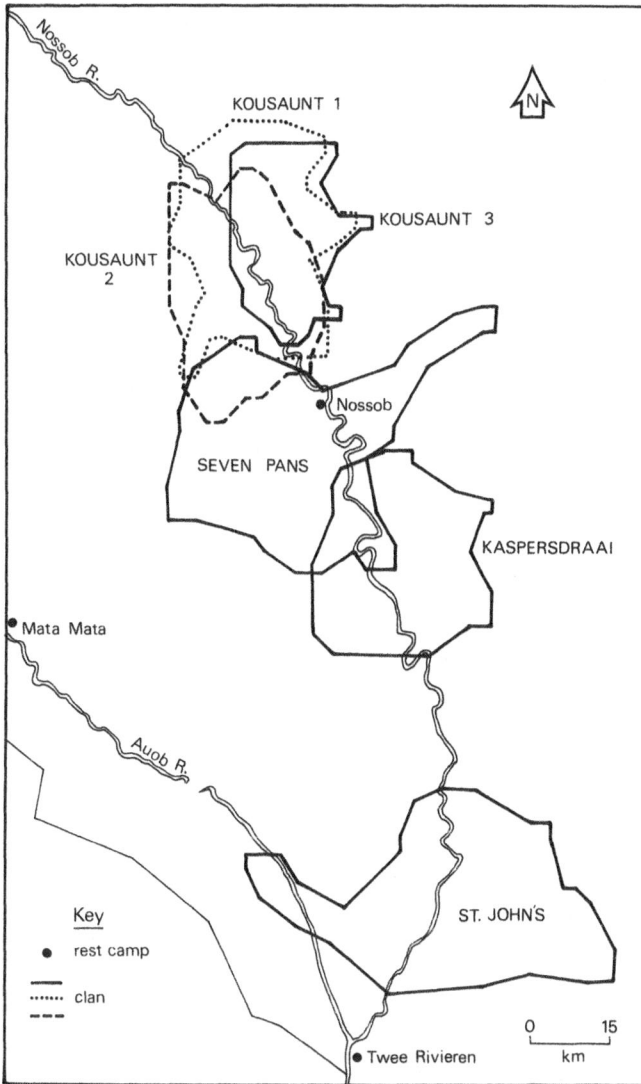

Figure 4.15 Restricted polygons enclosing the observed movements of spotted hyaenas from four clans. The Kousaunt clan is shown for three time periods: Kousaunt 1 1979–1980; Kousaunt 2 1981–1982; Kousaunt 3 1983.

My main objective in measuring territory size was to investigate the influence of the pattern of food dispersion on this component of social organization. The methods used to achieve this are discussed in Appendix D. There were considerable differences in territory size between the two species, and large fluctuations in territory size in both species, within and between clans. The mean size of brown hyaena territories (Fig. 4.14), measured by the restricted polygon method (Wolton 1985), and including the Kwang clan's territory over three time periods, was 308 ± 39 km^2 (Table 4.5). The mean of six spotted hyaena territories (Fig. 4.15) measured in the same way, and including the Kousaunt clan's territory over three time periods, was 1095 ± 177 km^2 (Table 4.5).

4.4.2   Factors affecting movement patterns

Studies of their movements within territories indicated that both species are affected by two factors: mainly the distribution of their food, but also the location of their dens.

As far as the distribution of food is concerned for both species, the density of ungulates is generally higher along the narrow river beds than in the extensive dunes (section 2.1.1, Table 4.6). Of importance to brown hyaenas are the facts that wild fruits grow almost exclusively in the dunes (Fig. 2.10), and several small animals occur at higher densities in the dunes, or at least in equal densities in the two habitats (Table 2.4). The effect of these food dispersion patterns on the movements of both species can be seen in differences in their relative use of dune and river habitats. Spotted hyaenas spent 55.2% of the distance they were followed foraging in river habitat, whereas brown hyaenas spent 45.2% of the distance followed (9). As the availability of their food shifted between these habitats, so too did the movements of the hyaenas. This is illustrated separately for the two species as follows:

1. Brown hyaena movement patterns varied depending on the relative amounts of river and dunes habitats in each territory. In a territory comprising mainly homogeneous dunes habitat where the food tends to be evenly distributed, such as the Seven Pans territory which comprised 95% dunes habitat (Fig. 4.14), the hyaenas' movements tended to be uniformly distributed within the territory (Fig. 4.16). In contrast, the movement patterns of brown hyaenas from the Kwang 2 territory (Fig. 4.14), which

Table 4.6   Average year round density of ungulates in ungulates per km^2 in river and dunes habitat in different spotted hyaena territories.

Territory	Springbok		Large herbivores	
	River	Dunes	River	Dunes
Kousaunt 1	23.7	-	5.3	-
Kousaunt 2	26.3	0.0	4.4	2.8
Kousaunt 3	29.6	0.2	4.3	2.2
Kaspersdraai	35.5	0.2	7.0	1.6
Seven Pans	25.3	0.4	9.1	1.8
St John's	28.6	0.0	4.7	0.2

comprised 37.8% river habitat, were far more clumped along the river bed (Fig. 4.17), showing their high utilization of this rich habitat.

2. The influence of river habitat on the overall movements of the Kousaunt spotted hyaena clan (Fig. 4.15) was marked. This is illustrated in Figure 4.18. However, as the ungulates moved between river bed and

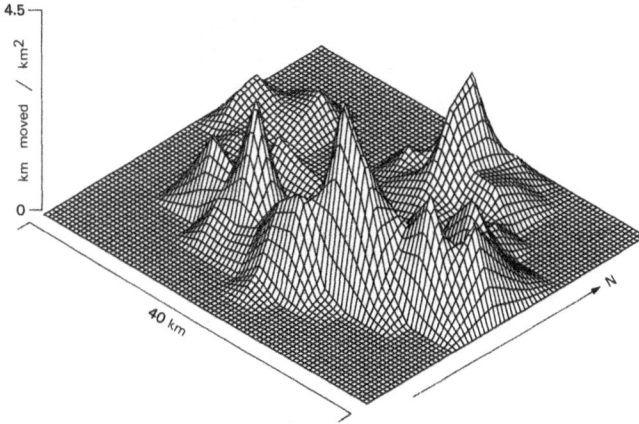

Figure 4.16 A three-dimensional map generated by Surface II (Sampson 1978) of the distances moved by brown hyaenas from the Seven Pans clan in different parts of the territory in 1976. The map was prepared directly from the matrix of distances moved in each of 6.25 km² map grids (see Appendix D) and is presented as viewed from the southeast, with the observation point situated 10 000 map units from the centre of the matrix and 35° above the horizontal.

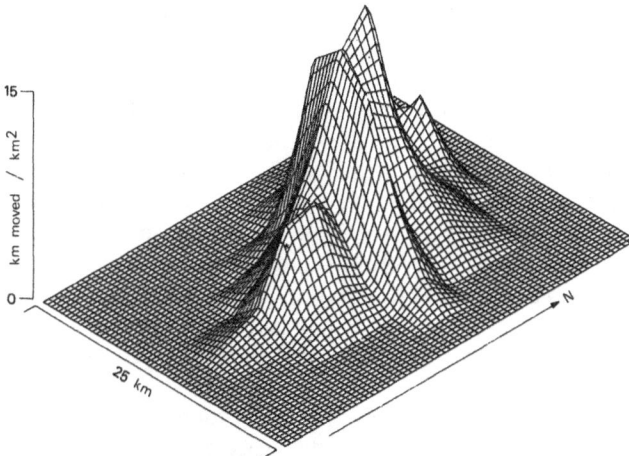

Figure 4.17 A three-dimensional map generated by Surface II (Sampson 1978) of the distances moved by brown hyaenas from the Kwang clan in different parts of the territory in 1976. Details as in Figure 4.16.

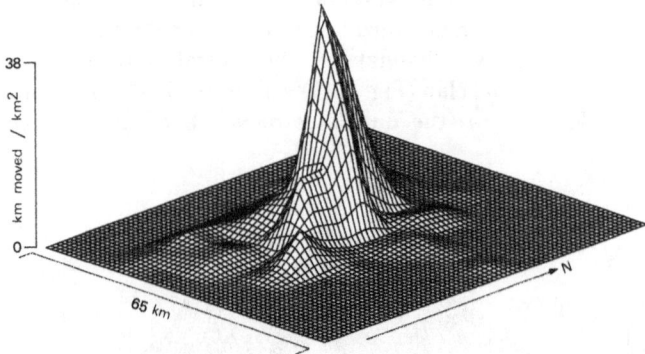

Figure 4.18 A three-dimensional map generated by Surface II (Sampson 1978) of the distances moved by spotted hyaenas from the Kousaunt clan in different parts of the territory for the entire study period. Details as in Figure 4.16, except that the map is viewed from 20° above the horizontal.

dunes, so spotted hyaenas' use of these two habitats changed. In 15 months when ungulate densities along the river bed in the Kousaunt territory were higher than the mean for the study in this habitat, the spotted hyaenas foraged more in river habitat than in the dunes. In 17 months when ungulate densities were lower than the mean along the river bed, the hyaenas foraged more in the dunes in ten months, and along the river bed in seven months. This is illustrated in Figures 4.19 and 4.20. Between November 1979 and January 1980, when herbivores were scarce along the river bed, but concentrated in the dunes, 45% of the distance moved by hyaenas was in river habitat, and 55% in dunes (Fig. 4.19). Between February and April 1980, when herbivores were concentrated along the river bed, but scarce in the dunes, 79% of their movements were in river habitat, and 21% in dunes (Fig. 4.20) (10). Clearly, the change in dispersion of prey influenced the hyaenas' movements.

In addition to the influence of the varying patterns of food availability, the movements of the hyaenas were affected by the locations of their breeding dens. The den is a focal point in hyaena society (section 6.1); females with cubs repeatedly return there to feed them, and other group members are also attracted to it. Often hyaenas will stay away from a den for a day or more, particularly if they have not found food, but, spotted hyaenas particularly, soon return to the den vicinity irrespective of their hunting success.

The location of dens (as with prey) was not random with respect to the two habitat types, and differed for the two species. Brown hyaenas located 94.4% of their dens in dunes habitat (Table 6.1), whereas spotted hyaenas often denned in river habitat (52.2% of their dens). The absence of brown hyaena dens from the river beds, therefore, is not because there are few

suitable dens there. As brown hyaena dens frequently have strong smelling
carcasses around them (a result of adults bringing food to the cubs), they
attract potential enemies, particularly spotted hyaenas (section 8.1.2). It
seems that brown hyaenas may locate their dens in the dunes in order to
keep away from the river beds, which are the main activity areas for
spotted hyaenas and also lions.

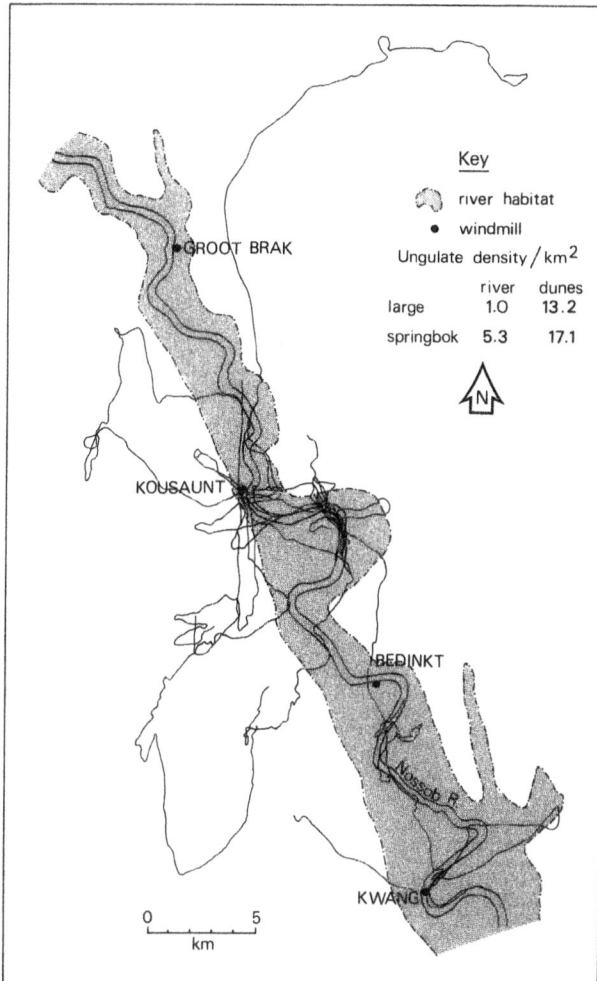

Figure 4.19 The observed movements of spotted hyaenas from the Kousaunt clan from
November 1979 to January 1980 and the density of ungulates at that time in river and dunes
habitat.

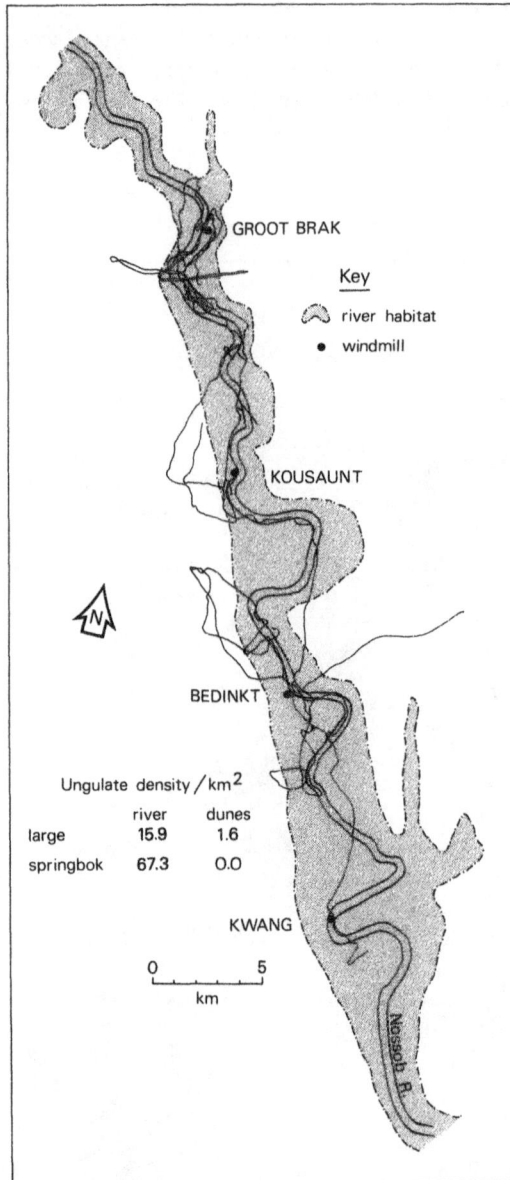

Figure 4.20 The observed movements of spotted hyaenas from the Kousaunt clan from February to April 1980 and the density of ungulates at that time in river and dunes habitat.

4.5   Factors affecting the sizes of social groups and territories

Recently, much evidence has been produced to support the hypothesis that patterns in the availability of resources, particularly of food, have been the main selective pressures facilitating group living in some animals. Amongst carnivores the advantages accruing to the members of these groups may be varied: it may enable them to increase hunting success and take larger prey through communal hunting (Kruuk 1975, Lamprecht 1978); larger groups may be more successful in dealing with both inter- and intraspecific competitors (Lamprecht 1981); they may be better able corporately to defend themselves against predators (Rood 1983, Rasa 1987); or some individuals may benefit through helping in the raising of offspring (Macdonald & Moehlman 1982). However, as pointed out by Carr & Macdonald (1986), a number of studies have suggested that group territoriality may be an epiphenomenon of the pattern of resource distribution, and that any advantanges of group living in these circumstances are secondary.

Bradbury & Vehrencamp (1976a, b) have constructed a model for the determination of what they call the 'social dispersion of organisms'; that is, the simultaneous comparison of group size and territory size. These, the model predicts, are limited by components of the food distribution. 'Territory size is determined by the average distance between successively available food sites, the number of such sites needed per year, and the average size of these sites. Maximal group size is then determined as a result of this territory size minimisation' (Bradbury & Vehrencamp 1976b: 383). This principle has been called the 'resource dispersion hypothesis' by Macdonald (1983), and states that territory size is determined by the dispersion of transient food patches, and group size is positively correlated with patch quality.

Support for this model has been found in studies of European badgers (Kruuk & Parish 1982 – although more recently Kruuk & Parish 1987) there have been some discrepancies between their findings and the resource dispersion hypothesis – red foxes (Macdonald 1981), and Arctic foxes (Hersteinsson & Macdonald 1982). The subject has been reviewed by Macdonald (1983), Kruuk & Macdonald (1985), and Carr & Macdonald (1986). Are the data from the present study compatible with this resource dispersion hypothesis?

In both species there were differences in group size and territory size, both within the same clan over time and between clans. However, as is shown in Figure 4.21, there was no correlation between the size of a clan's territory and the number of adults and sub-adults in that clan during the period that territory size was being measured. This suggests that these two parameters are affected by different variables. It also suggests that Kalahari hyaenas are not expansionists (Kruuk & Macdonald 1985); i.e. they do not form larger groups in order to expand their territories, and so increase their feeding efficiency, as has, for example, been suggested to be

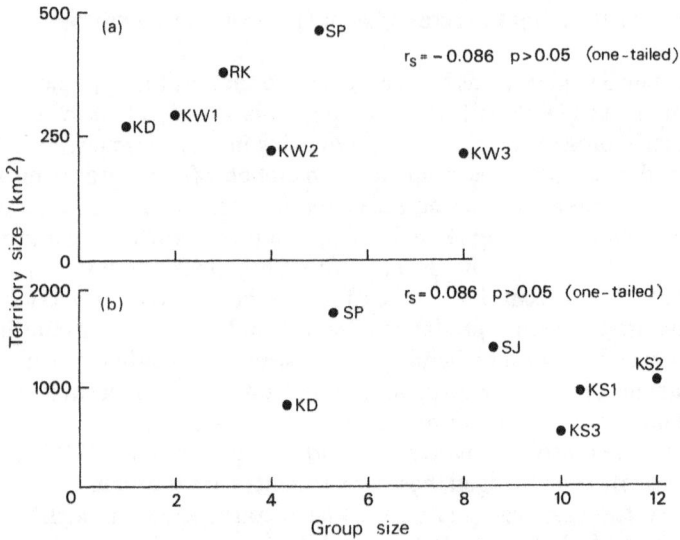

Figure 4.21 The relationship between territory size and clan size in (a) brown hyaenas and (b) spotted hyaenas. Abbreviations for clans are as follows: KD Kaspersdraai, KS1 Kousaunt 1979–80, KS2 Kousaunt 1981–82, KS3 Kousaunt 1983, KW1 Kwang 1975, KW2 Kwang 1976, KW3 Kwang 1977–78, RK Rooikop, SJ St John's, and SP Seven Pans.

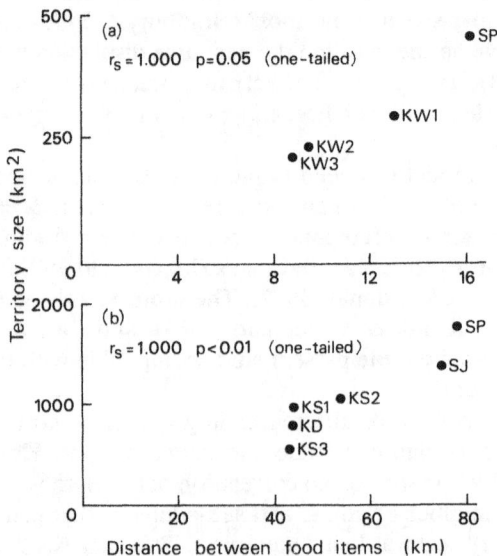

Figure 4.22 The relationship between territory size and the average distance moved between significant food items in (a) brown hyaenas and (b) spotted hyaenas. For brown hyaenas a significant meal was defined as a large vertebrate, or ten wild fruits. For spotted hyaenas it was a large or medium-sized kill, or meaty carcass. Abbreviations for clans as in Figure 4.21.

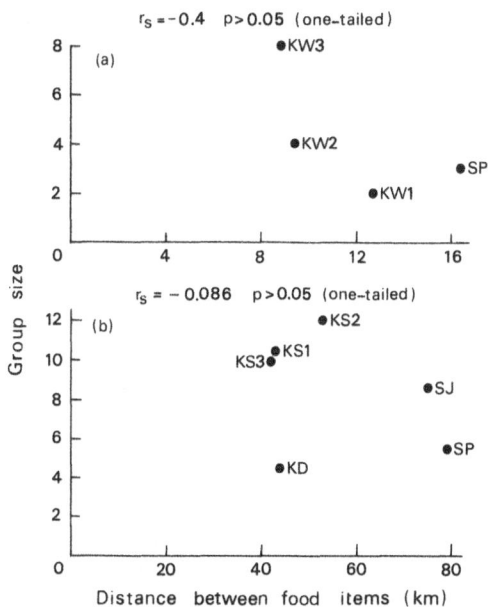

Figure 4.23 The relationship between clan size and the average distance moved between significant food items in (a) brown hyaenas and (b) spotted hyaenas. Definitions for significant meals are the same as in Figure 4.22, abbreviations for clans as in Figure 4.21.

the case for coyotes, by Bowen (1981, 1982). Nor do increases in group size in these hyaenas appear to require a proportional enlargement of territory, as Peterson *et al.* (1984) believed to be the case for wolves in Alaska.

Variation in territory size was not related to the distances for which hyaenas from each territory were followed (11). Therefore, the differences in territory sizes cannot be dismissed as sampling bias. For both species there was a significant relationship between the size of a territory and the average distance moved by its inhabitants between food items which were judged to provide a significant meal for at least one hyaena (Fig. 4.22). No such correlation existed between this parameter and group size (Fig. 4.23). These data, therefore, support the hypothesis that territory size is determined by the dispersion pattern of food, predominantly by the average distance between food sites, but that group size is not.

The manner in which the territory sizes of the Kwang brown hyaenas and the Kousaunt spotted hyaenas fluctuated suggests that Kalahari hyaenas do not follow the 'obstinate strategy' of a constant territory size in the model proposed by Von Schantz (1984). They should be regarded as contractors (Kruuk & Macdonald 1985), maintaining the smallest economically defensible area which will encompass sufficient resources for maintenance and reproduction, as predicted in the model of Bradbury & Vehrencamp (1976b).

According to the resource dispersion hypothesis, group size will be determined by the richness of the food patches, i.e. the amount of food available per patch in the territory. For carnivores this is a notoriously difficult parameter to measure accurately.

In Figure 4.24, the relationship between the size of the brown hyaena Kwang clan and aspects of food availability in the territory each year between 1975 and 1980 are compared. Only the mean number of wildebeest along the Nossob river bed in the territory each year correlated significantly with group size. A wildebeest, or other large herbivore carcass, can be regarded as a rich food patch for the generally solitary-feeding brown hyaena. It provides food for several individuals – the mean number feeding from a large carcass was 3.6 – although it need not necessarily feed the animals simultaneously. A smaller carcass, such as that of a springbok, usually only provided a meal for one or two brown hyaenas (mean 1.4). Moreover, there was evidence that in years when the average numbers of wildebeest and other large ungulates in the Kwang area were higher, more large ungulate carcasses became available to brown hyaenas than in other years (section 2.2.1). Patch richness here, therefore, may have been a function of the size of the carcasses which were available to the brown hyaenas and was correlated with group size.

I do not have the data to determine the relationship between brown hyaena territory size and patch richness, i.e. the number or biomass of any of the ungulates dying in the territory. This relationship is relevant, insofar as the prediction that group and territory sizes will not be correlated rests on the assumption that patch richness and patch dispersion (determining territory size) are not correlated. The indications are that no relationship existed. For example, the Kaspersdraai territory had few large and medium-sized ungulates in it (Table 2.2, where the data for 1972 were mainly from this territory), yet it was smaller than the Kwang 1 territory, but larger than the Kwang 2 and 3 territories (Table 4.5), all of which, because of the relatively large amounts of river habitat in them (Fig. 4.14), had higher large and medium-sized ungulate members.

Turning now to the spotted hyaena, it has been shown in section 4.2.1 that, with the exception of the Kaspersdraai, Seven Pans, and Urikaruus clans, there was little variation in clan size. Even over a six year period the Kousaunt clan contained 9–11 adults and sub-adults most of the time. With a coefficient of variation of 13.2%, the fluctuation in group size within the Kousaunt spotted hyaena clan over a six year period was considerably less than it was within the Kwang brown hyaena clan over the same length of time, where it was 41.9%. As the food eaten by spotted hyaenas in all these clans was similar, i.e. large and medium-sized ungulates, and because these occurred in more or less similar herd sizes in each of the territories for which I have data (Table 4.7), and almost certainly in the Leijersdraai and Kannagauss clan territories as well, patch richness did not appear to vary greatly from one territory to the other.

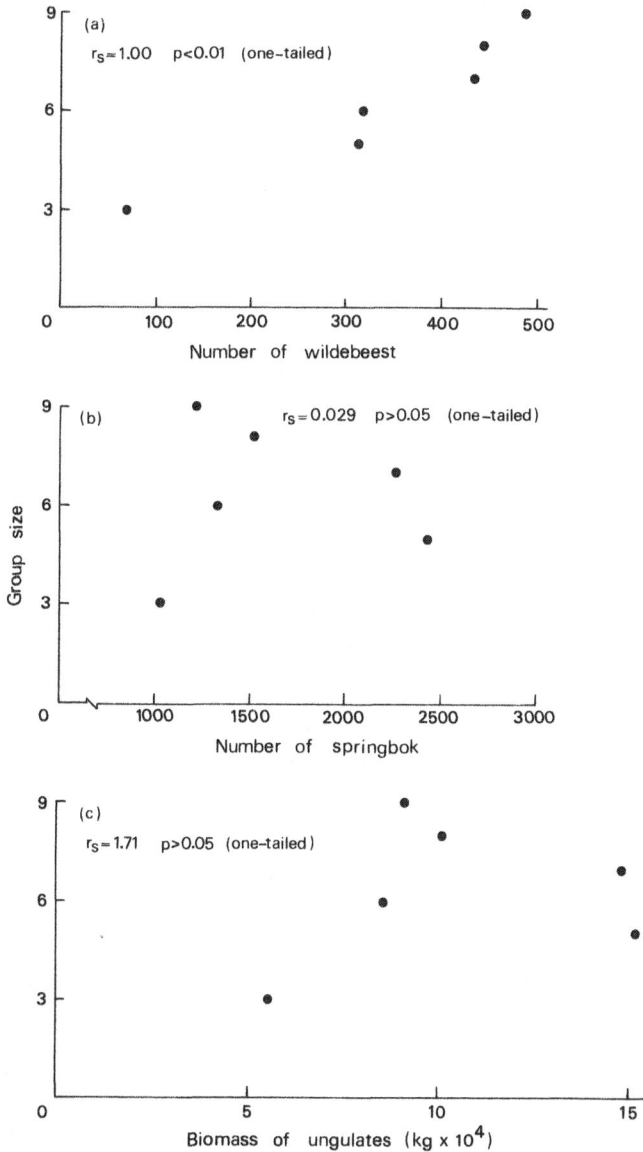

Figure 4.24 The relationship between brown hyaena clan size and certain aspects of food availability in the territory each year between 1975 and 1980. (a) The annual mean number of wildebeest counted along the river bed in the territory; (b) the annual mean number of springbok; and (c) the annual mean number of all ungulates converted to biomass.

Table 4.7   Mean ± SD herd sizes of gemsbok and wildebeest in different
spotted hyaena territories.

	Gemsbok	Wildebeest
Kousaunt	3.5 ± 5.3	7.2 ± 14.4
Kaspersdraai	4.1 ± 9.6	12.9 ± 23.0
Seven Pans	4.4 ± 6.3	14.3 ± 23.1
St. John's	3.0 ± 3.9	12.4 ± 22.9

Thus, in terms of the resource dispersion hypothesis clan sizes would be expected to be similar and to fluctuate little.

This hypothesis holds for most clans, but not for the three small clans. I know little about the Urikaruus clan, but there was evidence that the Kaspersdraai and Seven Pans clans were small for reasons not related to food. The Seven Pans clan may have been a recent breakaway group from another clan, as for at least three years prior to the appearance of this group in 1982, no spotted hyaenas occurred in the Seven Pans area. The Kaspersdraai clan was an established clan, which in mid-1980 had at least three breeding adult females plus other animals. However, by January 1981 this clan contained only two individuals, an adult female and an adult male. The reason for this sudden drop in numbers is unknown, but, in light of the observations reported earlier, could have been due to an outbreak of rabies. Be that as it may, however rich food patches may have been in these territories in 1983, it appears that these two clans could not have been larger in that year, because of limitations in breeding rate and the low frequency of female immigration. The Kaspersdraai clan was certainly growing; by July 1983 it consisted of five adults and sub-adults, and by January 1985 there were seven (Fig. 4.5b).

The evidence, therefore, suggests that for Kalahari spotted hyaenas territory size is determined by the manner in which the food resources are spaced out, and group size is at least partially influenced by the richness of the food resources in the territory, but other factors such as disease and social behaviour may mask this relationship.

Brown hyaenas in the southern Kalahari have a more catholic diet than do spotted hyaenas (Ch. 2). The average distance moved between significant food items (as defined in Fig. 4.22) for brown hyaenas was 7.2 km, compared with 32.7 km for spotted hyaenas. Spotted hyaenas need to travel longer distances between food items than do brown hyaenas, because of their more specialized feeding habits. Consequently, they need to forage over a much larger territory.

Although prey abundance in a spotted hyaena territory may vary (Table 4.6), their diet was not found to change markedly over time. This is in contrast to brown hyaenas where, for example in the Kwang area, their diet changed from predominantly medium-sized mammals to large mammals, as the latter became more available (Table 2.6). This led to a corresponding increase in group size (Fig. 4.1). Other things being equal, a homogeneous area like the southern Kalahari, where, apart from the river beds, food patches are few and ephemeral, fluctuations in group size would be expected to be greater in the more generalist brown hyaena than in the more specialized spotted hyaena. The higher coefficient of variation in the size of a brown hyaena clan in comparison with a spotted hyaena clan, is perhaps testimony to this. Brown hyaenas are able to capitalize on changes in food availability in the southern Kalahari more efficiently than are spotted hyaenas. Thus, although brown hyaenas are solitary foragers, whereas spotted hyaenas frequently forage in groups, clan sizes of the two species need not differ. A brown hyaena clan can be larger than a spotted hyaena clan. This, I suggest, is because social group size is determined by the quality of the resources in the territory, whereas foraging group size is determined by other selective pressures, such as enhanced efficiency in procuring food, the size of the food sources exploited, and possibly kin selection (see sections 3.2 and 7.3.3.2).

Data on group and territory size of brown hyaenas from other areas are few and somewhat confusing. Owens & Owens (1978, 1979a, b, 1984) studied a clan in the Central Kalahari Game Reserve, Botswana, which consisted of about 13 individuals of all ages, which is similar to the Kwang clan at its peak (Fig. 4.1). They suggested (Owens & Owens 1979b) that this was typical for the area, but did not give supporting data. In their 1978 paper they gave average individual dry season home range sizes, which they said were the maximum during the year, as $40 \, \text{km}^2$ (Owens & Owens 1978), but in a later paper (Owens & Owens 1979b) they gave average territory size as approximately $170 \, \text{km}^2$. It is not stated what method was used to measure these home ranges and territories, or whether the average territory size was calculated from the individual home ranges of members of the clan.

Goss (1986) studied a brown hyaena clan over a four month period near Luderitz along the coast in the Namib Desert. They fed mainly on a rich supply of seal pups, which apparently are available throughout the year. The clan consisted of a minimum of 12 individuals, and the clan's territory size of $220 \, \text{km}^2$ was measured by counting the number of $0.25 \, \text{km}^2$ grids through which the animals passed. This territory appeared to be excessively large as the bulk of the hyaenas' food was obtained on a $3 \, \text{km}$ stretch of coast. Goss (1986) hypothesized that a number of mining towns in the extreme inland portions of the territory, abandoned during the last 50 years, were once food sites which the hyaenas' ancestors had used and that the present-day hyaenas had not yet changed their lifestyle. If this is so, this is an unusually slow response by hyaenas to a change in food distribution patterns.

The variation in group and territory sizes of spotted hyaenas from different habitats is large (Table 4.8). The smallest Kalahari territory was 55 times larger than the largest measured territory in Hluhluwe Game Reserve, South Africa, and the smallest Ngorongoro clan was 2.5 times larger than the largest Kalahari one. The distribution of resources, in particular food, but in the Namib Desert water and shelter as well (Tilson & Henschel 1986), is believed to be the main reason for these differences. For example, in the Ngorongoro Crater spotted hyaenas live in larger clans and much smaller territories than in the southern Kalahari. Although they feed on similar sized prey animals in the two areas, the distance between food items is obviously less in the Ngorongoro, with its high density of prey (Kruuk 1972). Furthermore, herd sizes of prey are much larger in the Ngorongoro Crater than in the southern Kalahari, therefore food patches are richer.

The resource dispersion hypothesis, therefore, can account for much of the variability in group size and territory size in hyaenas. However, sometimes this principle may be overruled by other factors. I have suggested that in the southern Kalahari rabies and social behaviour may sometimes mask the relationship in spotted hyaenas, and Henschel (1986) obtained evidence that, as his Kruger National Park study clan's size decreased, a larger neighbouring clan appropriated part of their territory. Furthermore, Whateley & Brooks (1978) felt that group size should have been larger in Hluhluwe Game Reserve, considering the number of potential prey, and suggested that the rugged terrain and thick vegetation may have affected the hyaenas' hunting efficiency. Then there is the puzzling situation of brown hyaenas apparently inhabiting far larger territories than necessary along the Namib Desert coast (Goss 1986).

Table 4.8   Clan size and territory size of spotted hyaenas from different areas. Clan size refers to the number of animals older than nine months.

Area	n	Clan size	Territory size (km^2)	Source
Ngorongoro Crater	7	35 - 80	30 - 40	Kruuk 1972
Hluhluwe-Umfolozi Game Reserves	3	8 - 14	10 - 39	Whateley & Brooks 1978 Whateley 1981
Masai Mara National Reserve	1	52	60	Frank 1986a
Kruger National Park	1	11	130	Henschel 1986
Namib Desert	3	4 - 8	383 - 816	Tilson & Henschel 1986
Southern Kalahari	6	3 - 14	553 - 1776	This study

Similar fluctuations in group and territory size have been shown to occur in other carnivore species (see Macdonald 1983, and Kruuk & Macdonald 1985, for reviews). This widespread and extensive flexibility in social organization of carnivores, must be considered in comparative studies. It may place limitations on attempts to equate parameters such as home range size and metabolic needs of different carnivores, as done by Gittleman & Harvey (1982).

4.6 Summary

	Brown hyaena	Spotted hyaena
Mean and range of clan size	3.7 (1–10)	8 (3–12)
Coefficient of variation in clan size	42%	13%
Adult sex ratio in clans	Parity	Preponderance of females
Age structure of adults in the population	Mainly age-class 4	Mainly age-classes 2 and 3
Most important factors in clan dynamics	Inter-litter interval (mean 20.7 months) Male and female emigration	Inter-litter interval (mean 16.7 months) Male emigration Pack fission Rabies?
Clan size determined by	Food patch richness, i.e. the number of large ungulate carcasses	Food patch richness, i.e. mean herd size of important prey, possibly modified by rabies and social factors
Nomadic males	Important component of population 33% of adult males	Present, but form a smaller proportion of the population
Mean and range of territory size	308 km^2 (215–461 km^2)	1095 km^2 (552–1776 km^2)
Per cent distance moved in river habitat	44.8%	55.2%
Mean distance moved per meal	9.2 km	32.7 km
Territory size determined by	Distance between meals	Distance between meals

Statistical tests

1. Ratio of males:females in brown hyaena clans.
   Binomial test: $P = Q = 0.5$; $N = 19$; $x = 8$; $p = 0.648$; two-tailed.
2. Comparison of the frequency of brown hyaena litters born each month of the year versus the expected frequency if they were born equally throughout the year.

Kolmogorov–Smirnov one-sample test: $D = 0.19$; $N = 14$; $p > 0.05$; two-tailed.

3. Comparison of the ages at which brown hyaena males versus brown hyaena females disappeared from their natal clans.
    Mann–Whitney U test: $U = 12$; $n_1 = 3$; $n_2 = 9$; $p = 0.278$; two-tailed.

4. Ratio of males:females in spotted hyaena clans.
    Binomial test: $P = Q = 0.5$; $N = 21$; $x = 6$; $p = 0.039$; one-tailed.

5. Comparison of the frequency of spotted hyaena litters born each month of the year versus the expected frequency if they were born equally throughout the year.
    Kolmogorov–Smirnov one-sample test: $D = 0.1$; $N = 40$; $p > 0.05$; two-tailed.

6. Comparison of the observed frequency with which southern Kalahari adult spotted hyaenas from each age-class died versus the expected frequency if they died in proportion to which they were represented in the population.
    Kolmogorov–Smirnov one-sample test: $D = 0.36$; $N = 25$; $p < 0.01$; two-tailed.

7. Ratio of males:females of brown hyaenas resighted less than four times in known clans' territories.
    Binomial test: $P = Q = 0.5$; $N = 26$; $x = 9$; $p = 0.038$; one-tailed.

8. Ratio of males:females of brown hyaenas resighted four times or more in known clans' territories.
    Binomial test: $P = Q = 0.5$; $N = 18$; $x = 8$; $p = 0.407$; one-tailed.

9. Comparison of percentage movement over 10 km spent in river habitat for brown and spotted hyaenas.
    Mann–Whitney U test: $U = 1148$; $z = -1.929$; $n_1 = 35$; $n_2 = 40$; $p = 0.053$; two-tailed.

10. Comparison of the percentage of each movement that spotted hyaenas spent in river habitat between November 1979 and January 1980 versus the percentage between February and April 1980.
    Mann–Whitney U test: $U = 13.5$; $z = -3.23$; $n_1 = 11$; $n_2 = 12$; $p < 0.001$; two-tailed.

11. Correlation between territory size and the distance for which hyaenas from that territory were followed.
    (i)   *Brown hyaenas*. Spearman rank correlation coefficient: $r_s = 0.543$; $N = 4$; $p > 0.05$; one-tailed.
    (ii) *Spotted hyaenas*. Spearman rank correlation coefficient: $r_s = -0.257$; $N = 6$; $p > 0.05$; one-tailed.

5 Communication patterns and social interactions

Spotted hyaenas usually feed and forage together, brown hyaenas do so alone. This has influenced the evolution of many of their behaviour patterns. The spotted hyaena has a large number of group orientated behaviour patterns, whereas brown hyaenas tend to interact with conspecifics singly (Mills 1983a).

Order is maintained in animal societies by the transmission of information between individuals. In hyaenas this is achieved by visual and tactile signals, vocalizations, and scent marking (Mills *et al.* 1980, Gorman & Mills 1984, Mills & Gorman 1987). Interactions between individuals and a comparison of the ways in which the two species utilize these various communication patterns, as well as their behavioural flexibility, are discussed in this chapter.

5.1 Visual and tactile communication and social interactions

Visual and tactile signals, often augmented by vocal signals, are used by hyaenas to convey their mood and status to conspecifics. These form the basis of many social interactions.

5.1.1 Basic postures

The basic postures signalling a tendency to attack, to flee (submission), and appeasement are similar in the two species (Table 5.1). The main difference is that in the long-haired brown hyaena, pilo-erection is more prominent (Fig. 5.1) than in the short-haired spotted hyaena, and is displayed by both aggressive and submissive animals. Furthermore, in the spotted hyaena the dark tail contrasts with the light coloured body (Fig. 5.6), thus enhancing its signalling function, whereas in the brown hyaena both the tail and body are dark coloured (Fig. 5.1).

5.1.2 Territorial interactions

Hyaenas from different clans rarely meet up with each other in the southern Kalahari. In fact, I only observed one such meeting between spotted hyaenas, which was similar to the territorial conflicts described by Kruuk (1972). Several hyaenas from each clan chased each other back and

Table 5.1   The positions of certain body parts in both species, when displaying three basic postures.

	Tendency to attack	Tendency to flee (submission)	Appeasement
Stance	Erect	Crouched	Crouched, sometimes animal rolls onto its side
Hair	Erect	Erect in brown, down in spotted	Down
Head	High	Low	Low
Mouth	Closed	Open	Open with lips drawn back
Ears	Cocked	Flattened back	Flattened sideways
Tail	Elevated	Curled under body	Curled under body
Anal pouch	Often extruded	Not extruded	Not extruded
Forefeet	May paw the ground	No pawing	No pawing

Figure 5.1 A brown hyaena feeding on a springbok displays the tendency to attack posture at the approach of another.

forth, accompanied by much vocalizing, although in this instance no physical contact took place.

Several meetings between strange brown hyaenas were observed. In contrast to spotted hyaenas they were nearly always on a one-to-one basis, and were characterized by ritualized fighting involving biting at the neck. This behaviour has been called neck-biting by Owens & Owens (1978). An example:

10 September 1977. 03.00 h. A female brown hyaena from the Botswana clan arrives at a wildebeest carcass and starts eating. The carcass is situated in the area of overlap between the Botswana and Kwang territories along the Nossob river bed. A female from the Kwang clan is lying 25 m away. While the Botswana female feeds, the Kwang female raises her hair and starts soft growling (Table 5.6). After 15 min the Kwang female stands up and moves slowly towards the Botswana female with her head held low and her ears cocked. She stops for a few seconds some 10 m away, then runs over to the Botswana female, grabbing her on the side of the neck just below the ear, pulling her around. The Botswana female starts snarling and yelling (Table 5.6) with her mouth open and ears flattened backwards (Fig. 5.2). After about 1 min the Kwang female pulls her opponent down onto her stomach and stands over her, with her hair raised and ears cocked. The Botswana female lies, with her ears still flattened backwards and her mouth wide open, looking up at the Kwang animal. Keeping in these positions the two hyaenas pitch their heads from side to side with open mouths, but without making contact. The Botswana female is snarling and yelling all the time.

Figure 5.2   Two brown hyaena females from different clans neck-biting. The differences in the postures of the dominant (right) and submissive (left) animals can be clearly seen. Both animals are wearing radio-collars.

After 6 min the Botswana female stands up and moves away slowly, bleeding slightly from her left ear. The Kwang female stands, watching her go, and begins to paw the ground with alternate forefeet, hair raised and ears cocked. When the Botswana female is approximately 5 m away, the Kwang female chases her for 200 m, the latter running off out of sight.

I observed 16 neck-biting interactions lasting from a few seconds to several minutes. In 14 (88%) the contestants were not from the same clan, in two (12%) the origins of the contestants were unknown. Where I was able to sex the contestants, eight males and 13 females were involved, but animals of opposite sex rarely fought (see section 7.2.4). Where the position of a contestant in its territory was known, 11 (61%) were in boundary blocks and seven (39%) were internal blocks (see Fig. 5.12 for a definition of boundary and internal blocks), which, assuming that brown hyaenas spend 34.7% of their time in boundary blocks and 65.3% in internal blocks (Mills *et al.* 1980), suggests that most contestants were in boundary blocks (1). However, there was no increase in the likelihood of a hyaena winning a fight when in the interior of its territory than when on the boundary (2).

In all neck-biting interactions a winner clearly emerged. In five (31%) the winner chased after the loser, as in the example above, but in the remainder it merely watched the loser move off. On two occasions I followed the defeated animal after a fight in a boundary block, and in both instances it moved several kilometres into the interior of its territory. On six other occasions, when I stayed in the vicinity of the fight, the defeated animal was not seen again that night, even though there was a large carcass present. Only once, in the seven cases where I observed the winner's behaviour after a fight, did it immediately leave the vicinity, when it carried food back to its den. Thus, there was a marked tendency for the loser to retreat from the area and for the winner to stay (3).

These observations suggest that neck-biting has a territorial function, and are in contrast to those of Owens & Owens (1979b) in the central Kalahari, where, they maintain, neck-biting is carried out by members of the same group as an expression of dominance. They noticed no effort by the loser of a neck-biting bout to leave the area, and sometimes after neck-biting the two contestants would even feed side by side with no further interactions.

5.1.3   Interactions within clans

Interactions between clan members provide examples of different patterns of social interactions in the two species, depending on the degree and nature of sociality exhibited by each.

5.1.3.1 MEETING CEREMONIES

The meeting ceremonies of the two species are very different. When two brown hyaenas from the same clan meet they sniff each other around the head and face, along the neck and body, and/or at the anus. In meeting ceremonies involving anal sniffing, one animal usually presents its hind region to the other to be sniffed. The presenter cuts in front of the presentee with its head held low, its ears flattened out sideways, its lips drawn back into a 'grin', and often uttering a whine (see Table 5.6). It drops onto its carpals, so that the hind region is elevated and the tail is curled over the back, and extrudes the anal pouch, as the presentee sniffs it for a few seconds (Fig. 5.3). The roles may then be reversed.

Amongst brown hyaenas, cubs and sub-adults present to adults, but sub-adults and adults do not present to cubs, nor do adults present to sub-adults. When adults met, in about half the cases one presented to the other, and in about half the cases mutual or no presenting occurred. The same applied when sub-adults met (Table 5.2). Whether an animal is presented to, or whether mutual or no presenting occurs in a meeting ceremony between adults, does not appear to be sex-linked (4), nor were certain individuals more likely to present than others.

The posture adopted by the presenting brown hyaena during the meeting ceremony and the accompanying vocalizations are similar to those adopted by cubs when attempting to suckle (section 6.2.3). Such infantile-derived postures may cause a parental response rather than one which would be

Figure 5.3 A large brown hyaena cub presents its extruded anal gland to another during the meeting ceremony.

Table 5.2   Frequencies with which brown hyaenas of different age groups presented to each other during meeting ceremonies.

Age-groups	Frequency
Cub greeting adult or sub-adult	
Cub presents	Numerous
Adult or sub-adult presents	0 (0%)
Mutual or no presentation	0 (0%)
Sub-adult greeting adult	
Sub-adult presents	39 (98%)
Adult presents	0 (0%)
Mutual or no presentation	1 (2%)
Sub-adult greeting sub-adult	
One of the pair presents	1 (25%)
Mutual or no presentation	3 (75%)
Adult greeting adult	
One of the pair presents	14 (58%)
Mutual or no presentation	10 (42%)

directed at a rival (Ewer 1973). It is likely, therefore, that the meeting ceremony is derived from the begging behaviour of cubs and has an appeasement function, contributing to the maintenance of group cohesiveness.

The meeting ceremony of the spotted hyaena is far more elaborate than the brown hyaena's. It entails two hyaenas standing parallel to each other, heads in opposite directions, usually both, but sometimes only one of the animals lifting the hind leg nearest to its partner, and sniffing and sometimes licking the erect sexual organ of the other (Fig. 5.4). The behaviour is often accompanied by lowing and, in some cases, is preceded by one of the animals emitting a whine (see Table 5.6) and adopting an appeasement posture (Kruuk 1972).

In East Africa, meeting ceremonies occurred between spotted hyaenas of both sexes and of any age, being particularly frequent between cubs and adults (Kruuk 1972). However, I found that adult immigrant males rarely partook in meeting ceremonies, and hardly ever did so with adult females, although other categories of animals frequently performed the meeting ceremony with each other (Table 5.3).

On the two occasions that leg lifting occurred between an immigrant male spotted hyaena and a female, the male jumped away after a few seconds. On eight other occasions a male and female stood head to tail, but neither lifted a leg, the male again jumping away after a few seconds. The males were reluctant to present their genitals to the females. Presenting such a vulnerable part of the anatomy to a conspecific's most lethal

Figure 5.4 Female spotted hyaenas sniff each other's sexual organs during the meeting ceremony. The pseudoscrotum and erect peniform-clitorises can be clearly seen.

Table 5.3   Frequencies with which different categories of spotted hyaenas indulged in a full meeting ceremony with each other. The amount of time animals from different categories spent together was not recorded.

Category	Frequency
Immigrant male - immigrant male	0
Immigrant male - adult female	2
Immigrant male - sub-adult	12
Immigrant male - cub	7
Adult female - adult female	68
Adult female - sub-adult	58
Adult female - cub	48
Sub-adult - sub-adult	37
Sub-adult - cub	34
Cub - cub	7
Total	273

apparatus (Fig. 5.4) must entail a high degree of mutual confidence. This may explain why immigrant males, which are often excluded from intra-clan activities (section 7.3.2), partake so infrequently in meeting ceremonies with other clan members, particularly adult females.

As postulated for the brown hyaena, Kruuk (1972) thought the function of the meeting ceremony in the spotted hyaena was appeasement, providing a means whereby the social bonds between the members of a clan are re-established. As the members of a spotted hyaena clan need to cooperate, not only in hunting, but also in interacting with their major competitors, lions, and in coalitions within the clan (section 7.3.3.3), there would seem to be a need for a powerful bond-forming behaviour mechanism, such as the meeting ceremony.

Kruuk (1972) concluded that the enlarged genital structures found in female and cub spotted hyaenas provided a conspicuous structure for attention during the meeting ceremony. This viewpoint has been challenged. Gould & Vrba (1982) pointed out that the peniform clitoris and false scrotal sac arose through high foetal androgen levels, which are maintained in adult female spotted hyaenas (Racey & Skinner 1979, Lindeque & Skinner 1982b). They suggested that the high androgen levels evolved because of the need for spotted hyaena females to become larger than, and dominant to, males. This in turn led to the virilization of the external female sexual organs, which then were incorporated into the meeting ceremony. Hamilton et al. (1986) took an intermediate stance. They suggested that changes in female hormonal levels could have led to the initial rudimentary virilization of the female external genitalia. These were then incorporated into a genital display leading to further selection for the enlarged external genitalia found today in females. Recently, Van Jaarsveld & Skinner (1987) have suggested that the genital monomorphism in the spotted hyaena arose as a consequence of a punctuated genetic translocation, which resulted in female foetuses being exposed to raised levels of testosterone in females.

Whatever the origin of the enlarged genital structures of female and cub spotted hyaenas, it is clear that they are now incorporated into the meeting ceremony. Its frequent occurrence and elaborate nature suggest that it has become an important mechanism for maintaining the social bonds in a clan.

5.1.3.2 MUZZLE-WRESTLING AND PLAY

Muzzle-wrestling (Owens & Owens 1978) is important in intra-group interactions in the brown hyaena, but is absent in the spotted hyaena. The participants (usually two) stand face to face, sometimes dropping onto their carpals, and attempt to bite each other on the jowls, or the side of the neck, particularly at a white patch at the base of the ear (Fig. 5.5). They pitch their heads rapidly from side to side, as they parry and thrust at each other. If one manages to grab the other, it shakes it for a few seconds, or it

Figure 5.5 A sub-adult and cub brown hyaena muzzle-wrestle.

(usually the larger animal) may sometimes walk with the victim pulling it along for a short distance. Alternatively, it pulls its victim downwards, at which the wrestling continues with one standing over the other, or with both lying face to face. During the activity the contestants often soft growl (Table 5.6) and make a panting noise. In between clashes the animals may chase each other slowly, until the chased animal stops, often at a tree or bush, and the wrestling recommences. At other times between clashes the participants may lie down close to each other and groom themselves, or chew on some bones, or indulge in other unconnected activities. The duration of a muzzle-wrestling bout can be from a few minutes to over an hour, and it may take place anywhere in the territory. The vicinity of the den or of a large carcass is a favoured area. It usually occurs when the animals have fed well.

Animals from different age-groups indulged in muzzle-wrestling to different degrees (5). Sub-adults muzzle-wrestled more often than expected, and adults rarely muzzle-wrestled with each other, although they often did so with sub-adults (Table 5.4). Sub-adults, in fact, frequently attempted to initiate muzzle-wrestling with adults, although the adults did not readily respond, often rebuffing the sub-adult, or moving away from it.

Muzzle-wrestling has elements in common with neck-biting, but the clear difference between winner and loser in neck-biting is not discernible and the contestants remain together, or drift away slowly after the bout. Neck-biting is a more aggressive behaviour pattern than is muzzle-wrestling.

Table 5.4   Analysis of muzzle-wrestling bouts between brown hyaenas of different ages from the Kwang clan. A bout is defined as a sequence of muzzle-wrestling during which the participants do not move out of sight of each other or engage in some other activity.

Age-groups	Number of hyaena hours observations	Number of muzzle-wrestling bouts observed	Muzzle-wrestling bouts per hyaena hour	Expected number of muzzle-wrestling bouts
Cub - cub	330	33	0.10	28.4
Cub - sub-adult	157	22	0.14	13.5
Cub - adult	316	11	0.03	27.2
Sub adult - sub-adult	31	6	0.19	2.7
Sub adult - adult	140	23	0.16	12.1
Adult - adult	152	2	0.01	13.1
Total	1126	97	-	97.0

However, on occasions muzzle-wrestling escalated into an aggressive encounter. This was occasionally so when sub-adults muzzle-wrestled with each other, and, more strikingly, when certain adult females muzzle-wrestled with sub-adult females (section 7.2.3). In the latter cases the younger animals displayed some of the characteristics of the losing animal in neck-biting; they flattened their ears backwards, opened their mouths and made little attempt to retaliate. They also uttered various whining sounds which are characteristic of submissiveness (Table 5.6).

The closest behaviour pattern to muzzle-wrestling found in the spotted hyaena is what Kruuk (1972) described as social play. It is usually indulged in by several animals which chase and bite each other, the chased hyaena frequently carrying a bone or a stick, the pursuer(s) either trying to bite the quarry in the hind legs, or to get hold of the object it is carrying. Twice a spotted hyaena went to a bush, broke off a twig and then elicited play with others by running past them with the twig. As with muzzle-wrestling in brown hyaenas, most play in spotted hyaenas took place between cubs and sub-adults (Table 5.5).

The resemblance of the brown hyaena's muzzle-wrestling to the spotted hyaena's play, and the fact that young animals were so often involved in this behaviour, suggests that play is a function of muzzle-wrestling. Bekoff & Byers (1985) (and references therein) have suggested that the presence of certain motor patterns along with the vigorous exercise exhibited in play, may be important in the development of these particular motor skills, as well as in general physical fitness. This could be particularly important for young animals of both species.

Bekoff & Byers (1985) have also suggested that play may serve to form and maintain social bonds. When muzzle-wrestling with adults, sub-adults may be attempting to integrate themselves into the clan. Muzzle-wrestling bouts between adults and sub-adults were always initiated by the sub-

Table 5.5   Frequencies with which different categories of
spotted hyaenas played with each other. The
amount of time animals from different categories
spent together was not recorded.

Category	Frequency
Immigrant male - immigrant male	0
Immigrant male - adult female	0
Immigrant male - sub-adult	1
Immigrant male - cub	1
Adult female - adult female	2
Adult female - sub-adult	9
Adult female - cub	9
Sub-adult - sub-adult	14
Sub-adult - cub	26
Cub - cub	23
Total	85

adults, which sometimes went to great lengths to get the adults to react. When adult females and sub-adult females muzzle-wrestled it may have been an expression of competition for a breeding place in the clan (section 7.2.3). When sub-adults muzzle-wrestled with each other they too may have been competing for a place in the social group, particularly when the interactions became more aggressive than normal. In spotted hyaenas the maintenance of social bonds may be achieved more effectively by other behaviour patterns, such as the meeting ceremony and social sniffing (sections 5.1.3.1, 5.1.3.3), thus making play a less important behaviour in this species when compared with muzzle-wrestling in the brown hyaena.

5.1.3.3   SOCIAL SNIFFING
Social sniffing, first described by Kruuk (1972), is only found in the spotted hyaena. Several hyaenas sniff the ground excitedly with their tails up, their bodies touching and making fast head movements up and down. Some lie on their briskets, others remain standing. Kruuk (1972), found nothing uniting the places where spotted hyaenas social sniffed, although on occasions they chose a place where another spotted hyaena had been lying, or a latrine or pasting place. He concluded that the function of social sniffing is not related to scent, but rather the 'doing of something together' as an aid to social integration.

Social sniffing may be an important aid to social integration, but I got the impression that the stimulus for it often appeared to be the smell of strange hyaenas. Of 17 social sniffing bouts observed, nine took place at latrines,

Figure 5.6 Three spotted hyaenas harass a fourth (note its ear sticking out between the two animals on the left) in a behaviour similar to Kruuk's (1972) female baiting. Note the prominant dark tails of the animals.

three at a patch of urine, one at a pasting place, one at a grass clump which smelt of hyaena, and only three at places which had no scent detectable to me.

5.1.3.4 FEMALE BAITING

Another group-orientated behaviour found only in the spotted hyaena is female-baiting (Kruuk 1972). The baited animal takes up a crouched position, ears flat and teeth bared, while the baiters sniff or bite it (Fig. 5.6). The interaction is accompanied by snarling, lowing, yelling, fast whooping, and giggling (see Table 5.6).

The participants in the southern Kalahari differed somewhat from those in East Africa. Of ten baiting interactions observed, nine were directed at an adult female, but one was directed at a sub-adult male. Furthermore, in seven of the interactions some of the baiters were females, whereas in East Africa they were always males only.

Female baiting occurred when an animal arrived at a place where several others were lying, but the baited animal was not necessarily the last arrival. This behaviour may be related to dominance relationships between individuals, or coalitions (section 7.2.4).

5.1.4 Functional considerations

The basic postures used to signal a tendency to attack, flee or appease, are similar in the two species. These, no doubt, bear testimony to their close phylogenetic relations. On the other hand, there are marked interspecific differences in some visual and tactile communication patterns and social interactions:

• Most spotted hyaena interactions are group-orientated, whereas brown hyaenas usually interact in dyads.
• The two ritualized postural behaviour patterns, neck-biting and muzzle-wrestling, are found only in the brown hyaena. (Although elements of both are found in the spotted hyaena, they are too weakly developed to be able to classify interactions between individuals on this basis.)
• The spotted hyaena has evolved a complex meeting ceremony, different from the usual carnivore greeting pattern of anal sniffing (Ewer 1973) – the mode of greeting adapted by the brown hyaena and the other members of the Hyaenidae (Kruuk 1976, Richardson 1985, this study).
• Spotted hyaenas have more conspicuous tails than brown hyaenas, but
• Pilo-erection is a conspicuous posture in the brown hyaena (as well as the striped hyaena and aardwolf), not in the spotted hyaena, which only erects the short hairs on the neck.

What accounts for these fundamental differences? As spotted hyaenas spend much time in each other's company, group-orientated interactions would be expected to evolve. This may explain the lack of behaviour patterns such as neck-biting and muzzle-wrestling in their behavioural repertoire – these should be present in species where dyadic interactions are common, such as the brown hyaena. The most obvious dyadic behaviour pattern in the spotted hyaena is the unique meeting ceremony, which can be seen as a manifestation of a complex social system. Finally, the increased signalling efficiency of the spotted hyaena's tail may make it more efficient in transmitting messages in the confusion of communal interactions, whereas pilo-erection may be more pronounced in the smaller members of the Hyaenidae to make them look larger in conflict situations.

5.2   Vocalizations

Most vocalizations of hyaenas grade into related vocalizations; few are discrete. This increases the scope for subtle communication, but complicates the task of categorizing the calls. A preliminary comparison of the vocal systems of the two species, and discussion on the only long-distance vocalization found in the Hyaenidae, the whoop of the spotted hyaena, are presented in this section.

5.2.1   A comparison of the vocal repertoires of the brown hyaena and spotted hyaena

In conjunction with G. Peters and J. Henschel (personal communication), preliminary classifications of the vocal repertoires of both species have

Table 5.6   The vocal repertoires of the brown hyaena and spotted hyaena.

Vocalization	Kruuk's 1972 term	Description	Brown hyaena	Spotted hyaena	Situation
Whoop	Whoop	See section 5.2.2	-	+	See section 5.2.2
Fast whoop	Fast whoop	See section 5.2.2	-	+	See section 5.2.2
Low	Groan and low	Ooo-sound of variable pitch and loudness	-	+	By dominant during the meeting ceremony and during aggressive intra- and interspecific inter-actions. Also heard around dens usually by females with cubs.
Soft growl	Grunt	Low pitched, soft, deep-throated growl	+	+	By females to cubs, also in agonistic situations such as two strangers meeting. With brown hyaena, also sometimes during muzzle-wrestling.
Snarl	Growl	High frequency, gutteral sound	+	+	In defensive posture, when being threatened or approached by dominant, in both intra- and inter-specific interactions.
Short snarl	-	Low pitched, loud gutteral sound of short duration	+	-	In interspecific clashes, especially when standing its ground to spotted hyaenas.
Roar-growl	-	Loud, explosive growl	-	+	When being bitten.
Rumble	Soft grunt-laugh	Series of soft, rapid staccarto grunts	-	+	Alarm call, typically when several are feeding on a carcass. All scatter.
Laugh-grunt	-	Short, medium amplitude sound of variable pitch	+	+	During play-wrestling and low intensity agonistic situations in both species.

Hoot-laugh	Loud grunt-laugh	Soft sound, consisting of several low, but variable pitched, tonal-staccato elements	+	+	Accompanied by tendency to attack posture. In spotted hyaena often uttered by several animals simultaneously.
Giggle	Giggle	Loud, high pitched staccato sound	+	-	By submissive animal when being attacked or chased by a con-specific, particularly at food.
Yell	Yell	Loud of varying pitch, starting low and working into a crescendo	+	+	By an animal being bitten. In the brown hyaena emitted in conjunction with the snarl by the loser during neck-biting.
Squeal	-	High pitched, tonal sound of medium to high amplitude	+	-	By submissive animals when being attacked.
Harsh whine	Whine	Loud, high-pitched, staccato squeal ee-ee-ee-ee	+	+	With appeasement posture; beg call of cubs wanting to suckle and in brown hyaena also uttered by submissive adults during agonistic encounters.
Whine	Soft squeal	Softer, less staccato squeal	+	+	With appeasement posture by cubs and sometimes submissive adults during the meeting ceremony. In brown hyaena also by submissive animal in low intensity agonistic situations and occasionally during allogrooming.

been made. These, together with descriptions of the vocalizations, and the situations in which they were heard, are given in Table 5.6. Eight vocalizations have been identified in the brown hyaena, although Owens & Owens (1978) recognized only five. All the brown hyaena's vocalizations are short-distance vocalizations (although the yell can be heard over several kilometres), and mostly occur in dyadic interactions. The hoot-laugh is the only discrete vocalization.

The vocal repertoire of the spotted hyaena has been described by Kruuk (1972), and Henschel (1986). Modifying Kruuk's system, 14 vocalizations have been identified for this species (Table 5.6). They occur in both dyadic and communal interactions, often comprising several sounds in combination with each other, making them particularly difficult to unravel. The hoot-laugh and the rumble are the only discrete calls.

The spotted hyaena not only has a larger vocal repertoire than the brown hyaena, but also uses it more frequently, more loudly, and with more variation. The vocalizations found in the spotted hyaena only, such as the whoop, the low, the rumble, the giggle, and the squeal, are important in communal antagonistic situations such as intra- and inter-clan fights, and mobbing lions, i.e. situations in which brown hyaenas do not find themselves. Some of the vocalizations made by female spotted hyaenas to cubs, particularly the low and soft growl, are also heard in agonistic situations. These may be bond-facilitating vocalizations rather than aggressive.

5.2.2 The whoop call

The whoop is one of the characteristic sounds of the African night and is extensively used by spotted hyaenas, yet little is known of its function. Kruuk (1972) recorded that the whoop was usually uttered by a single animal and appeared to be spontaneous. Hyaenas rarely 'answered' each other's whoops.

5.2.2.1 WHOOPING AND RESPONSES TO THE CALL

Henschel (1986) described how the whoop starts off as a deep lowing sound (phase 1), followed by a slow rise in pitch (phase 2), rising to a high (phase 3). This is sometimes followed by a descending low (phase 4) which may then occasionally rise again (phase 5). Each whoop lasts 2–5 s and 5–9 whoops are usually made during each bout. Individuals have characteristic whoops identifiable by other hyaenas. There is tonal variation in different individuals' whoops, and the number of phases may vary. For example, the male, Hans, only had a phase 4 component in 2% of his whoops ($N = 154$), whereas the male, Nicholas, had a phase 4 component in 94% of his ($N = 48$).

Immigrant males whoop more often than members of other social classes, adult females whoop more often than sub-adults, and sub-adult males whoop more often than sub-adult females (Fig. 5.7 and Table 5.7).

In agreement with Kruuk (1972), situations under which spotted hyaenas whooped (Table 5.8) suggested that most (60%) whooping was spontaneous. On three of the occasions that whooping appeared to be employed in locating other clan members (Table 5.8), the hyaenas whooped after becoming separated and soon found each other. However, the best example of this use of whooping is the one described in section 2.6.2.1, during an encounter over food with a lioness, when a hyaena whooped 17 times and was soon joined by three others.

The observed responses of hyaenas to whoops (Table 5.9), show that those out of sight of the whooper were more likely to look up (and presumably to listen more attentively) than those in its presence. These data show a similar frequency as in Table 5.8 with which individuals

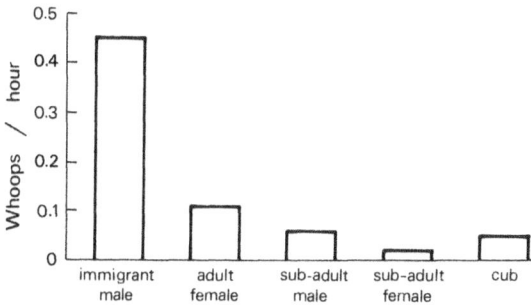

Figure 5.7 The mean rates at which spotted hyaenas from different social categories whooped.

Table 5.7   One-tailed Mann-Whitney U tests comparing differences in whooping frequency for different categories of spotted hyaenas.

	Immigrant male $n = 2$	Adult female $n = 7$	Sub-adult male $n = 12$	Sub-adult female $n = 4$
Immigrant male	-	$U = 0.0$ $p = 0.036$	$U = 0.0$ $p = 0.025$	$U = 2.5$ $p < 0.05$
Adult female		-	$U = 4.5$ $p < 0.015$	$U = 0.0$ $p = 0.018$
Sub-adult male			-	$U = 2.54$ $p < 0.05$
Sub-adult female				-

Table 5.8   Situations and/or stimuli during which spotted hyaenas whooped (excluding communal interspecific encounters).

Situation	Frequency
No obvious stimulus	150 (60.0%)
Within one minute of another whooping	46 (18.4%)
Prior to or on moving off	31 (12.4%)
Attempting to locate other clan members	13 (5.2%)
After an unsuccessful hunt of a large animal	5 (2.0%)
By an immigrant male at the approach of another immigrant male	3 (1.2%)
While scraping the ground with forefeet after a hyaena from another clan had whooped close by	2 (0.8%)
Total	250

Table 5.9   Responses of spotted hyaenas to other's whoops (excluding communal interspecific encounters).

Response	Whooper observed	Whooper not observed
No visible reaction	71 (67.6%)	39 (31.0%)
Whoop ('answer')	19 (18.1%)	27 (21.4%)
Lift heads and 'listen'	1 (1.0%)	37 (29.4%)
Move off in direction of whoop	-	22 (17.5%)
Move off with whooper	14 (13.3%)	-
Turn around and move off in opposite direction	-	1 (0.8%)
Total	105	126

'answered' whoops, and a similar association between whooping and the commencement of a movement. They also support the hypothesis that hyaenas sometimes whoop to locate each other, as in 17.5% of the cases the responder moved off in the direction of the whooper. So do eight observations of a hyaena coming up to another within 5 min of the latter whooping.

Of the 126 cases in Table 5.9 that a hyaena out of sight whooped, only two individuals were known to be members of a different clan from the one I was observing. These two observations are recounted below:

17 July 1981. Three adult females and the immigrant male of the Kousaunt clan have spent the day in the far western region of their territory. At sunset a hyaena whoops 1–2 km to the west of them. The hyaenas stand up immediately and look off in that direction, heads held

high, ears cocked. One of the females whoops and then the male does, scratching the ground vigorously with his forefeet. The four hyaenas circle around for two minutes, tails up and lowing softly, then lie down again. Ten minutes later they move off in an easterly direction.

2 July 1982. 01.55 h. Nine members of the Kousaunt clan are moving in a southerly direction near Seven Pans, in the area of overlap between the Kousaunt and Seven Pans clans (see Fig. 4.15). As they come to a latrine a hyaena whoops about 1 km away to the south. The hyaenas look in that direction for several seconds, then start lowing softly, tails up. Several of them paste (see section 5.3.1), defecate and scratch at the latrine, before moving back in the direction from which they had come. After 1 km one of them whoops as they continue moving deeper into their territory.

To test hyaenas' reactions to the whoops of members of other clans I played tape-recordings of both strangers and clan members to certain hyaenas. The tapes were played when the hyaenas were in a relaxed state, either lying at their dens, or eating peacefully on a large carcass near the centre of their territory. Hyaenas took little notice of the whoops of fellow clan members, but reacted aggressively to the whoops of strangers (Table 5.10).

A variation on the whoop is the fast whoop (Kruuk 1972), which is higher pitched with the calls and intervals between them shorter than in the whoop. Kruuk (1972) recorded that the fast whoop was mainly used in confrontations with lions or hyaenas from other clans. Table 5.11 records the situations during which I observed fast whooping. Because of the low density of lions and spotted hyaenas in the study area, interspecific clashes with lions and inter-clan battles were of rare occurrence. Whenever they did take place whooping and fast whooping were heard. The relatively high incidence of fast whooping during meeting ceremonies occurred on some occasions when one hyaena, usually a sub-adult, came up to several others, and a measure of aggression similar to female baiting (section 5.1.3.4) was exhibited by the others. The noise that the individual concerned made (fast whooping, squealing, whining) seemed out of proportion to the attention it received.

Table 5.10   Reactions of spotted hyaenas to tape recordings of the whoops of other spotted hyaenas.

	Take little notice	Move aggressively towards source of sound
Whoop of fellow clan member	4	0
Whoop of hyaena from another clan	3	8

Fisher exact probability test p = 0.026 (one-tailed)

Table 5.11   Situations during which spotted hyaenas were observed to fast whoop.

Situation	Frequency observed
During meeting ceremonies	15 (54%)
During encounters with lions	10 (36%)
During agonistic encounters between clan members not involving food	2 (7%)
During encounters with hyaenas from other clans	1 (4%)
Total	28

5.2.2.2   FUNCTIONAL CONSIDERATIONS

The whoop is used in several contexts, and, therefore, probably has several functions. The whoops of hyaenas from other clans elicit aggressive reactions in the residents, and whooping is also heard in other aggressive situations, indicating that it may be a mechanism in territorial maintenance. Much of the 'passive' whooping by hyaenas may be to advertise their presence in the area to intruders.

Territorial fights between hyaenas are rare because individuals avoid each other. Even when neighbours do meet the conflict is usually settled without fighting, when the intruder withdraws (section 5.1.2). For territorial conflicts to be settled by such a convention it is necessary that the resident animals can be unambiguously identified. One way that this may be achieved in the spotted hyaena is through whooping. Intruders entering a territory will hear the owners whooping and will perceive that the area is occupied. Further advance might lead to an encounter with the residents who would probably be prepared to fight ferociously. Should intruders meet up with residents the whooping which takes place during these encounters will identify them as such (individuals recognize each other's whoops), and the conflict can be settled conventionally. Vocalizations, therefore, may act as a cue to establish an asymmetry of contest, allowing for a conventional settlement and preventing escalation to real fighting (Maynard Smith & Parker 1976). (Scent marking is believed to function in a similar manner (section 5.3.4).)

The fact that adult immigrant males whoop more often than do other social classes of hyaenas may indicate a need for these animals to emphasize their presence to the adult females in the area, and to warn off other prospective immigrant males (see section 7.3.2.3).

Another function of the whoop may be to enable spotted hyaenas to locate each other. However, why hyaenas should sometimes react by seeking the whooper, when as far as I could discern the whoops were no different from those which did not elicit such a response, is unclear. It may depend on the motivational state of the animals concerned, e.g. hunger, or on the preceding events, e.g. becoming separated when foraging. Nor is it clear why spotted hyaenas should sometimes whoop when they start moving.

The fast whoop is aggressively motivated and appears to be a rallying call. It is usually united with other aggressive vocalizations such as the hoot-laugh and low (Table 5.6), and is a communal vocalization.

5.3 Scent marking

Hyaenas scent-mark in various ways. The secretion from anal glands of strong-smelling substances on to grass stalks (Figs 5.8, 5.10) is unique to the Hyaenidae and has been termed pasting (Kruuk 1972, 1976, Kruuk & Sands 1972, Mills *et al.* 1980). They also regularly defecate at latrines, where accumulations of conspicuous white faeces collect (Figs 5.15, 5.18). Associated with both pasting and defecation at latrines, and also occasionally seen on its own, is vigorous scraping of the ground with the forepaws, leaving discernible marks, and, in the case of the spotted hyaena at least, secretions from interdigital glands (Kruuk 1972).

5.3.1 Scent marking behaviour and comparisons in the structure of the scent pouches of the two species

26 May 1976. 04.00 h. An adult male brown hyaena from the Kwang clan is in the south of the territory. He comes onto a vehicle track and moves along it. After 100 m he stops and moves over to a grass stalk on the side of the road. He sniffs at the grass stalk, moving his nose up and down it with his ears cocked and his tail up, then moves forward, lifting a foreleg over the stalk and turning slightly as he does so. In this way the grass stalk runs forward under his belly and the base of the grass comes to lie between his hind legs. With his tail curled up over his back and his back legs slightly bent, he extrudes his anal pouch. For several seconds he feels for the grass stalk and eventually succeeds in locating it in the groove running down the white central area of the anal pouch (Fig. 5.8). He moves forward, pulling the anal pouch along the grass stalk and at the same time retracting it. The first effect of this action is to smear a thick, creamy blob of white paste on to the grass stem. Then, as the pouch continues to retract, a thin smear of black secretion is deposited some distance above the blob of white paste (Fig. 5.9). The hyaena continues forward, walking over the grass stalk, which springs back into its original position as he moves on.

Three hundred metres further on he comes to an accumulation of faeces on the side of the track. He sniffs at several of the faeces then defecates, after which he vigorously scratches the sand with his forepaws in the following sequence: two right, four left, seven right, five left, six right. He moves away and 50 m further on stops to sniff at another grass stalk, which has several pastings on it. He pastes on top and continues. He moves another 0.5 km down the track, during which he pastes four more times, and defecates and scratches at another latrine.

This is a typical sequence of scent marking by a brown hyaena when near the boundary of its territory. Spotted hyaenas scent-mark in a similar manner, but they are usually in a group when they do so. They also usually paste on a clump of grass stalks (Fig. 5.10) rather than a single one, and usually do so with the sexual organs erect.

The anal pouch lies between the rectum and the base of the tail. In the brown hyaena it consists of two distinct regions (Fig. 5.8); a large central area, which is normally covered in an accumulation of white secretion, and, lying one to each side of it, two circular areas which produce the black secretion. The two regions are separated from each other by non-secretory epithelium. The brown hyaena is unique among the Hyaenidae in that it secretes two pastes. The smell of the white paste can still be detected by the

Figure 5.8 A brown hyaena deposits a paste on a grass stalk on which there is another pasting. The large central area of the anal gland which secretes the white paste, with groove running down the centre, and small lateral area which secretes the black paste, can be clearly seen.

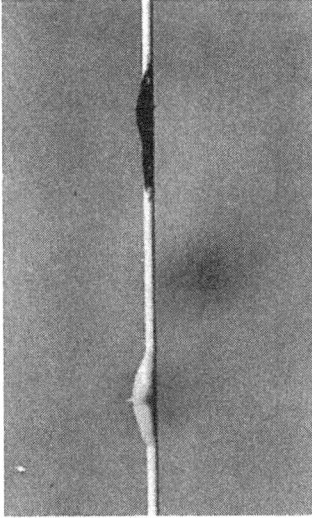

Figure 5.9 A pasting of a brown hyaena showing the two substances secreted.

Figure 5.10 A spotted hyaena pastes on a grass clump.

human nose well over 30 days after deposition, by which stage it has turned black. The smell of the black paste is not as long lasting, and is indiscernible to the human nose after a few hours.

The anal pouch of the spotted hyaena consists of only one region (Fig. 5.11), and is smaller than the brown hyaena's. The paste produced is creamy-white, and the odour lasts for at least as long as the white paste of the brown hyaena. The pastes of the two species smell quite different; the brown hyaena's has a 'salty' odour, whereas the spotted hyaena's has a more pungent smell.

At the histological level, the central area of the pouch which produces the white secretion in the brown hyaena is composed of numerous enlarged sebaceous glands, and the white paste is rich in lipid (Mills *et al.* 1980). The circular areas which produce the black secretion consist almost entirely of apocrine sudoriferous tissue. The black colour of the paste is due to accumulations of lipo-fuchsin, a common metabolite of apocrine tissue, and contains little lipid (Mills *et al.* 1980). In the spotted hyaena the major secretory elements are a pair of enlarged and lobulated sebaceous glands which open into the pouch via two large ducts (Fig. 5.11) (Matthews 1939, M. Gorman personal communication). As in the brown hyaena, this sebaceous tissue and the paste produced by it are rich in lipid (M. Gorman personal communication).

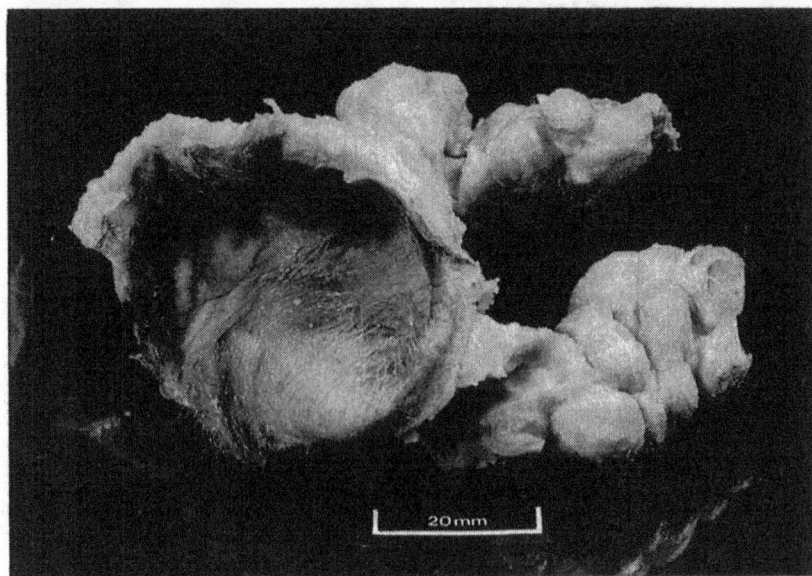

Figure 5.11 The dissected-out anal gland of a spotted hyaena showing the pair of lobulated sebaceous glands leading into the anal pouch. (Photo M. Gorman).

5.3.2 Factors affecting the distribution and abundance of scent marks

5.3.2.1 BROWN HYAENA

The overall pasting frequency of seven brown hyaenas followed for 1947 km was 2.64 pastings/km, with no difference between the pasting frequencies of males and females (6). The number of active scent marks in a territory at any given time is a function of the rates at which new marks are deposited and old ones decay. In the Kwang territory between 1976 and 1978 there were five to seven adult and sub-adult animals, each travelling approximately 30 km/night (section 3.1), and pasting at a rate of 2.64 marks/km. Taking 30 days as a conservative estimate of the average length of life of a scent mark, at any given time there would have been 15 000 or more active pastings in the territory.

Figure 5.12 shows the distribution of pastings (and latrines) I saw deposited in the Kwang 2 (1976) territory, showing how they were scattered throughout the whole territory. Although this only represents a small proportion of the total number of marks that were deposited, it is a representative sample, as the hyaenas were followed for long periods of time, often for two or more complete nights. This sample was subjected to a nearest-neighbour analysis and compared to the expected distribution if the locations were randomly distributed (Patterson 1965). This analysis showed that the pastings were more regularly spaced than random through the territory (7). Although some pastings were deposited at latrines the majority were not situated near any obvious landmark.

The spatial distribution of the sample of pasting sites is presented as a three-dimensional map in Figure 5.13. The figure shows that although pastings were found throughout the territory, the highest densities were near to the centre with a progressive decrease towards the borders. The reason for this is that the hyaenas spent more time in the central area of the territory (Fig. 4.17), and the number of pastings deposited per grid block was proportional to the distance moved in that block (8). However, the pasting frequency in boundary blocks (mean = 4.46/km) was higher than in internal blocks (mean = 2.31/km) (9). On three occasions brown hyaenas went well outside the clan's territory. During these excursions the rate of pasting dropped markedly to 0.4/km. During one particularly long excursion of 17 km, the hyaena sniffed at numerous pastings, but did not paste at all.

Figure 5.12 also shows the positions of latrines observed to be used by the Kwang 2 clan. Latrines were also scattered throughout the territory, but again not uniformly so. A nearest-neighbour analysis of all the latrine locations within the territory revealed that their distribution was more clumped than random (10). The groupings of latrines occurred mainly near the border of the territory, particularly in the north and south where it crossed the Nossob river bed.

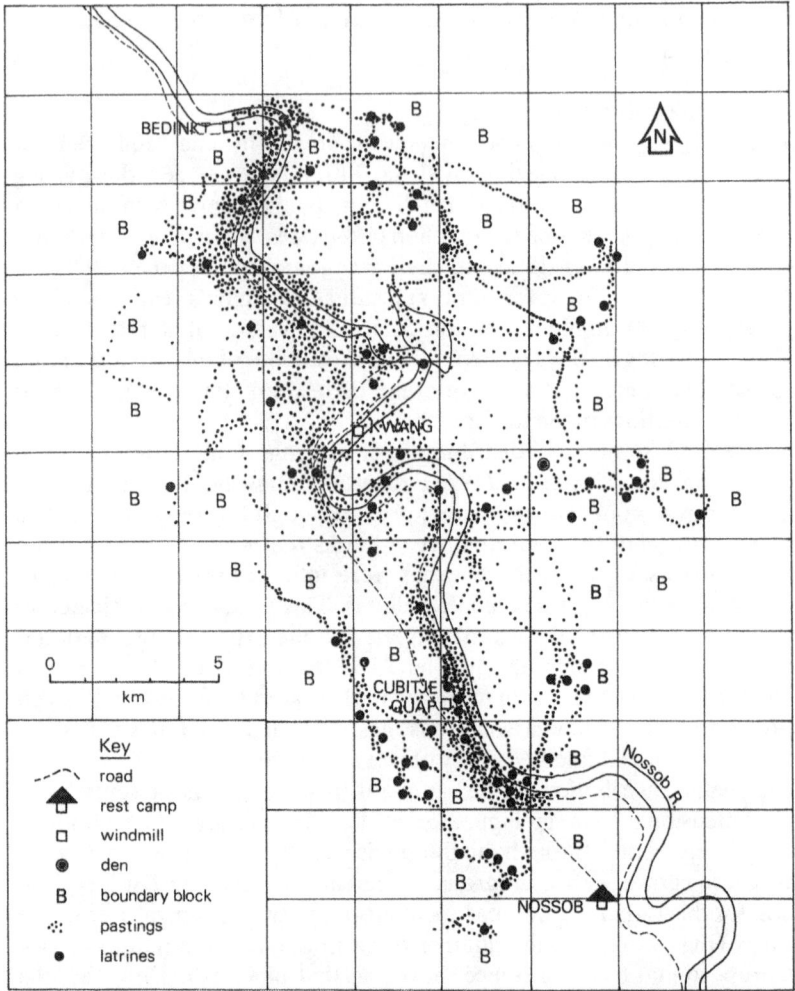

Figure 5.12 A map of the locations at which brown hyaenas of the Kwang clan were seen to paste and visit latrines in 1976. The 6.25 km² grid system is shown, with those grids on the periphery of the territory designated boundary blocks and the rest internal blocks (Mills *et al.* 1980).

Most brown hyaena latrines were situated next to a landmark (Fig. 5.14), mainly shepherd's trees (Fig. 5.15), which are the most common tree in the dunes. There was a marked tendency for latrines to be clumped on the south of these trees (11). At 34 randomly chosen latrines situated at shepherd's trees the latrines were on the south side of the tree in 24 (71%) and on the west side in seven (21%), with the rest being on the east side.

The frequency with which brown hyaenas visited latrines, expressed as

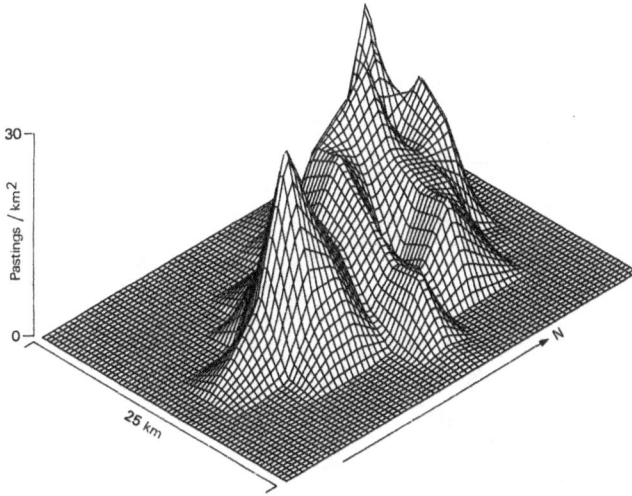

Figure 5.13 A three-dimensional map generated by Surface II (Sampson 1978) of the density of pastings in the Kwang brown hyaena territory in 1976. The map was prepared directly from the matrix of pastings deposited in each of the 6.25 km² map grids, and is presented as viewed from the southeast with the observation points situated 10 000 map units from the centre of the matrix and 35° above the horizontal.

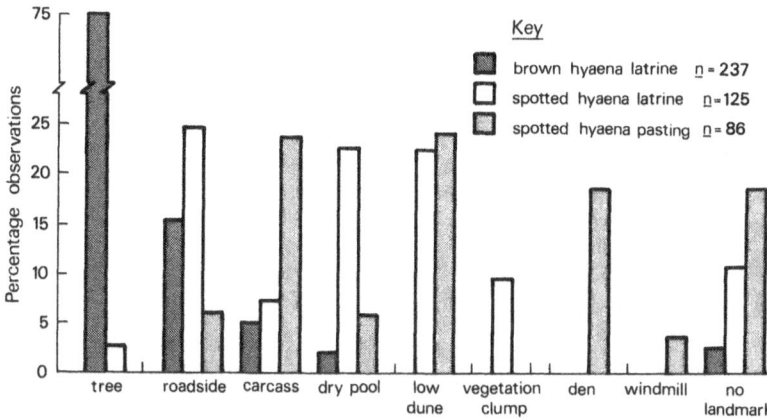

Figure 5.14 The proportions of brown hyaena latrines and spotted hyaena latrines, and pasting sites associated with various environmental features.

latrines visited per kilometre travelled, was also significantly higher when hyaenas were in boundary blocks (mean = 0.24) than when in internal blocks (mean = 0.07) (12).

Of 159 latrines visited by brown hyaenas of the Kwang clan, 128 (80.5%) were observed to be visited only once. This suggests that the number of

Figure 5.15   A brown hyaena latrine situated next to a shepherd's tree.

latrines found represents a small proportion of the total number in a territory. However, as I have argued for pastings, they are probably representative of the real distribution of latrines in the territory. It also suggests, as was borne out by observations on some specific latrines, that many are only used over a short period, after which they are abandoned. Such temporary latrines might be near a large carcass which would attract several hyaenas for a few nights, after which they would cease to visit the area. Other latrines are more permanent, for example, one that I found in August 1972 was still being used in January 1984. Similar temporary and long-term latrines have been recorded in spotted hyaenas by Bearder & Randall (1978).

The response of a brown hyaena to a pasting depends largely on where in the territory it is encountered. It will be more likely to over-paste it if the pasting is in a border block than if it is in an internal one (Table 5.12). As a paste in a border square has a higher chance of belonging to a hyaena from another clan than does one in an internal block, it may be that brown hyaenas paste on top of foreign pastes, but not on those from fellow group members. Observations of hyaenas which came upon a fresh pasting whose author was known (either because I had taken the pasting from one territory and placed it in another, or because I had seen a hyaena deposit the paste earlier), strongly suggest that brown hyaenas can distinguish between pastings belonging to their own clan and those deposited by foreigners. They are more likely to investigate closely and paste on top of foreigners' pastings than on those from their own clan (Table 5.13). Mills *et*

Table 5.12   The behaviour of brown hyaenas towards pastings encountered in
              different parts of the territory.

	Border square	Internal square	χ^2
Paste on top of paste or close by	65	26	34.31; d.f. = 1; $p < 0.001$
Did not paste	23 .	61	

Table 5.13   The reactions of brown hyaenas to pastings from their own and foreign clans.

	Own clan's pasting	Foreign pasting	Fisher exact probability
Approach to within 1m, but do not sniff or paste	6	0	$p = 0.0092$
Approach to within 1m then sniff and/or paste	6	11	

al. (1980) have shown that the complex chemical compositions of the
pastings of individual brown hyaenas are different in terms of the relative
concentrations of their chemical components, and have suggested that
these differences can be used by hyaenas to identify the author of a pasting.

5.3.2.2 SPOTTED HYAENA

The overall pasting frequency of 24 spotted hyaenas followed for over
8000 hyaena km was 0.13 pastings/km – a value 20 times lower than the
frequency for brown hyaenas. Given that there were on average 11 adults
and sub-adults in the Kousaunt clan, that each travelled about 27.0 km/
night (section 3.1), and pasted at a frequency of 0.13/km, and that a pasting
has an effective life of 30 days, there would have been about 1160 active
pastings in the territory at any one time. This is considerably less than the
15 000 pastings calculated for the brown hyaena Kwang territory.

Figure 5.16 shows the locations of all the scent marking sites I saw the
members of the Kousaunt clan use. These sites consisted of both latrines,
where the hyaenas defecated and pasted, and sites at which they only
pasted. Scratching the ground may have taken place at either type of site,
but was not systematically noted.

The frequency distribution of the number of pastings deposited at
latrines was quite different to that for sites receiving pastings only (Fig.

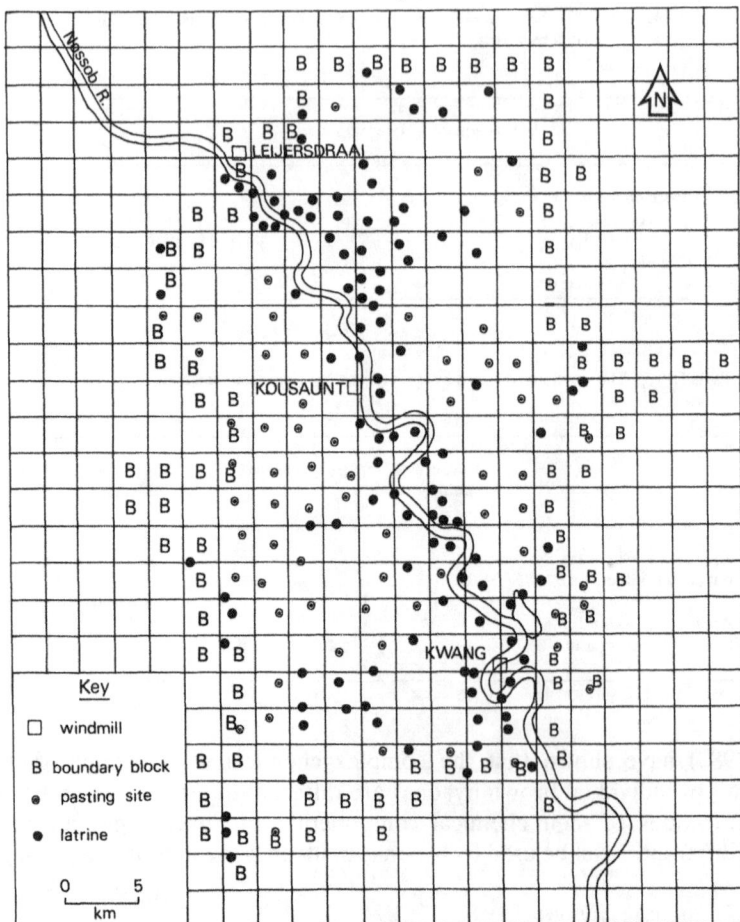

Figure 5.16 A map of the locations at which spotted hyaenas of the Kousaunt clan were seen to paste and visit latrines. As in Figure 5.13, the 6.25 km² grid system is shown, with boundary blocks marked.

5.17). Pasting sites usually received fewer pastings than did latrines, although some sites, particularly kills, received large numbers.

Both latrines and pasting sites were scattered throughout the territory, but not uniformly so (Fig. 5.16). Nearest neighbour analyses of the latrines, and of the individual pastings (not pasting sites), revealed that both were significantly more clumped than random (13, 14). In comparison, brown hyaena latrines were also more clumped than random, whereas pastings were more regularly spaced than random.

Latrines and other pasting sites were distributed differently. The sites receiving pastes only were generally dispersed throughout the territory,

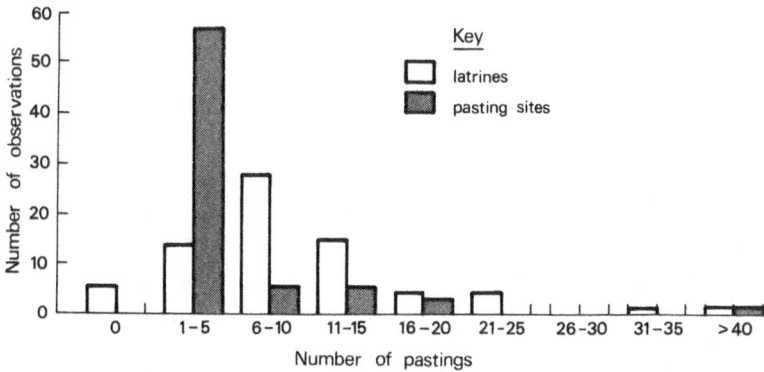

Figure 5.17 The frequency distributions of the number of pastings seen to be deposited by spotted hyaenas at latrines and other pasting sites.

whereas, as with brown hyaenas, latrines tended to be concentrated around the river bed, particularly towards the northern and southern parts of the territory (Fig. 5.16). However, internal map blocks contained more latrines (mean = 0.91) than border blocks (mean = 0.034) (15).

The majority of pastings deposited were placed either at latrines (63%), or at some other obvious environmental feature, although there were differences between the landmarks used for latrines and for pure pasting points (Fig. 5.14). Brown hyaenas and spotted hyaenas showed a marked tendency to use different sites for latrines (Fig. 5.14); only 2.4% of the spotted hyaena latrines were placed at shepherd's trees, compared with 66% of brown hyaena latrines, and spotted hyaenas used dried out pools (Fig. 5.18) and the tops of small dunes to a far greater extent than did brown hyaenas.

Spatial differences in the density of pastings throughout the Kousaunt territory, based on the number deposited in each map block, are shown in Figure 5.19. As with brown hyaenas, the highest densities of pastings were found in the interior of the territory and not at the periphery. Also, as was the case with brown hyaenas, this difference in the density of pastings between the two regions of the territory is due to the fact that the hyaenas utilized the interior to a greater extent than they did the border (Fig. 4.18), although, contrary to brown hyaenas, there was no significant correlation between the total distance travelled in any given square and the number of pastes deposited in that square (16). This can be seen when comparing Figures 4.18, 5.19.

The locations of pastings deposited by adults and sub-adults differed greatly. Sub-adults were far more likely to paste at a den (26.8% of pastings) than were adults (1.1%), this difference not being due to differences in the amount of time the two age-groups spent at dens. In contrast to brown hyaenas, there were also differences in the rates at which

Figure 5.18   A spotted hyaena latrine situated next to a dried out pool.

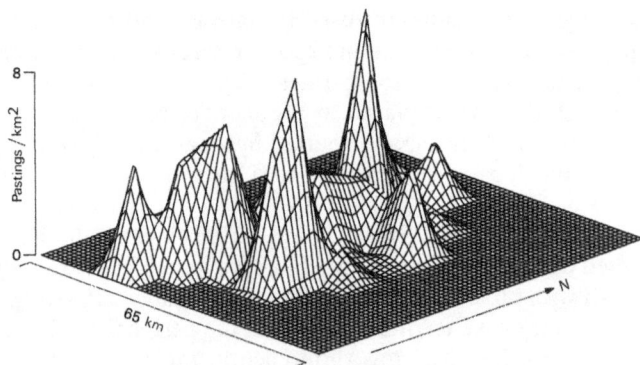

Figure 5.19   A three-dimensional map generated by Surface II (Sampson 1978) of the density of pastings in the Kousaunt spotted hyaena territory in 1976. The map was prepared as in Figure 5.13, except that it is viewed from 20° above the horizontal.

different social categories of spotted hyaenas pasted (Fig. 5.20 and Table 5.14) and defecated at latrines (Fig. 5.20 and Table 5.15), adult immigrant males doing so at the highest frequencies. Immigrant males were also more likely to scratch the ground with their forepaws after pasting or defecating.

The tendency, so prevalent in brown hyaenas, for animals to paste at higher frequencies in boundary blocks than in internal blocks was not

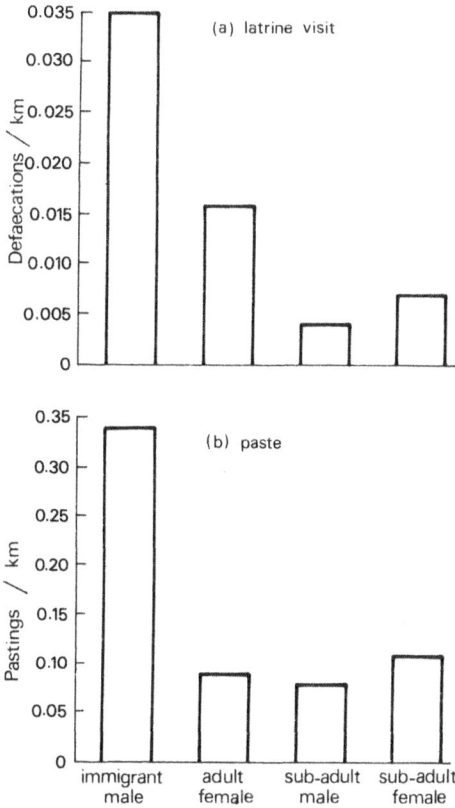

Figure 5.20 The mean rates at which spotted hyaenas from different social categories (a) defecated at latrines and (b) pasted.

Table 5.14   One-tailed Mann-Whitney U tests comparing the rates at which spotted hyaenas of different social categories pasted.

	Immigrant male $n=2$	Adult female $n=7$	Sub-adult male $n=12$	Sub-adult female $n=4$
Immigrant male	-	$U = 0.0; p = 0.028$	$U = 0.0; p = 0.017$	$U = 0.0; p = 0.067$
Adult female		-	$U = 33.0; p > 0.05$	$U = 11.0; p > 0.05$
Sub-adult male			-	$U = 13.5; p > 0.05$
Sub-adult female				-

Table 5.15 One-tailed Mann-Whitney U tests comparing the rates at which spotted hyaenas of different social categories visited latrines.

	Immigrant male $n = 2$	Adult female $n = 7$	Sub-adult male $n = 12$	Sub-adult female $n = 4$
Immigrant male	-	$U = 0.0; p = 0.028$	$U = 0.0; p < 0.001$	$U = 0.0; p = 0.067$
Adult female		-	$U = 19.5; p = 0.05$	$U = 12.0; p > 0.05$
Sub-adult male			-	$U = 12.0; p > 0.05$
Sub-adult female				-

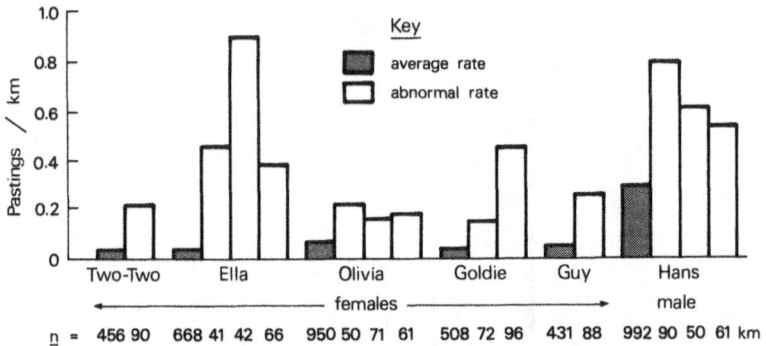

Figure 5.21 Details of journeys on which spotted hyaenas pasted at abnormally high rates and the mean frequencies with which these individuals were observed to paste. The distances that the animals were observed to move in each case are given below each bar.

found in spotted hyaenas. The average pasting frequency in boundary blocks was 0.14 compared to 0.12 in internal blocks (17).

On four foraging expeditions involving six adults and several sub-adults and cubs, individuals pasted at much higher rates than usual. The pasting frequencies for each adult on each expedition in which it took part, together with the average rates for those individuals, are shown in Figure 5.21. These 'abnormal' bouts of pasting were associated with long distance movements to the periphery of the territory. At certain latrines the hyaenas sniffed repeatedly at grass stalks, often indulging in social sniffing (section 5.1.3.3), and pasting extensively; for example, two hyaenas pasted 17 times at one latrine and then 11 times at another only 0.8 km away. It is possible that other hyaenas had recently visited these latrines as at one of them fresh faeces were seen and at another strange hyaenas were heard whooping close by. Therefore, like brown hyaenas, spotted hyaenas may be able to distinguish between the pastes of individuals, and react by over-pasting foreign pastes more intensively than they do the pastes of fellow clan members.

5.3.3   The efficiency of scent marking

Computer simulations were used to measure the efficiency of scent marking, by asking how likely it is that a scent mark will be encountered by an intruder. A 'hyaena' was placed to an accuracy of 1 m at several thousand random positions within simulated maps of a brown hyaena and spotted hyaena territory. For each random position the distance between the 'hyaena' and the nearest pasting site, given the density and dispersion patterns of scent marks for the two species, was measured. From these minimum distances, contours of equal value using Surface II were generated for each species (Figs 5.22, 5.23). The analyses show that a brown hyaena will on average be 500 m from a scent mark (the 500 m contour in Fig. 5.22 contains 99% of the scent marks), and over most of the territory it will be within 250 m. A spotted hyaena will be 1–2 km from the nearest scent mark over most of the territory. In the inner part of the territory it

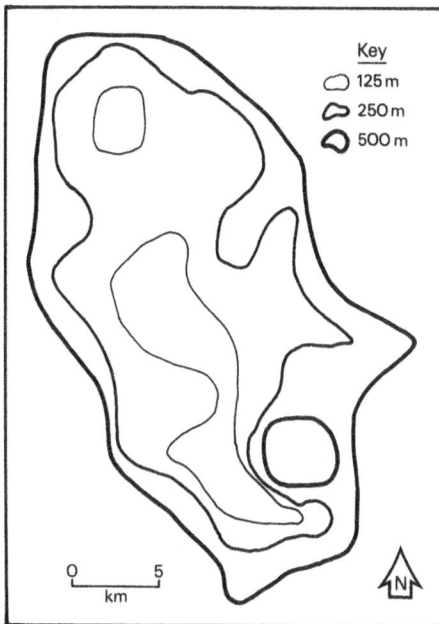

Figure 5.22   A contour map of the average distance between a randomly positioned brown hyaena and the nearest scent mark belonging to the Kwang clan. The 500 m contour approximates to the territory border.

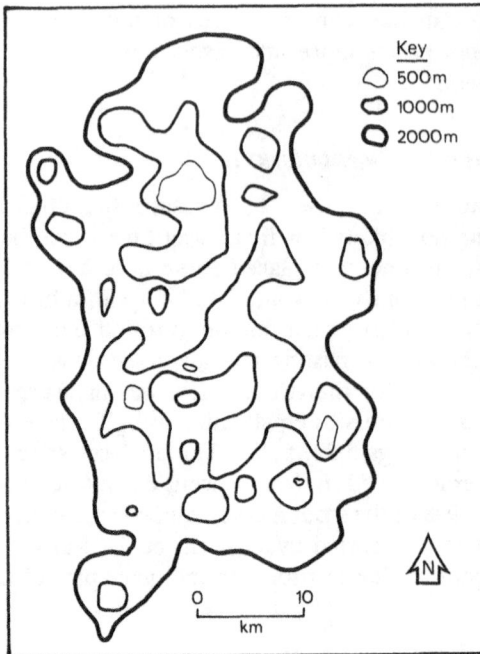

Figure 5.23 A contour map of the average distance between a randomly positioned spotted hyaena and the nearest scent mark site of the Kousaunt clan. The 2 000m contour approximates to the territory border.

will be within 500 m (Fig. 5.23). Since these analyses are based on a small sample of the actual number of scent marks present, these distances will probably be lower in reality. Given the efficient sense of smell of hyaenas and the strong odours emanating from the pastes, it is clear that intruders of both species entering a territory would encounter marks soon after their transgression.

5.3.4   Functional considerations

Southern Kalahari hyaenas employ a hinterland scent-marking strategy, placing most marks in the areas they utilize most frequently. This is different from the border marking strategy found in spotted hyaenas in the Ngorongoro Crater (Kruuk 1972), and in aardwolves (Richardson 1985).

However, brown hyaenas paste at a far higher frequency when near the boundaries of their territories, and their latrines are likely to be close to boundaries, as was also found to be the case with spotted hyaenas in the Kruger National Park (Henschel 1986). Furthermore, brown hyaenas respond to the presence of foreign marks by overmarking them. Southern Kalahari spotted hyaenas do not paste at a higher frequency when in boundary blocks – nevertheless, they did show abnormally high rates of scent marking during a number of long-distance forays that took them near to the border of their territory. There was evidence that these high rates of pasting were associated with the presence of marks deposited by the members of neighbouring clans. Their latrines too were often associated with areas where the chances of contact with strange hyaenas may have been highest.

The scent-marking behaviour of both species is consistent with Gosling's (1982) hypothesis that scent marks function to allow the unambiguous recognition of those animals which have invested heavily in an area, and which are likely to defend it by all-out fighting. This is a similar function to the one I hypothesized for the spotted hyaena's whoop call (section 5.2.2.2). The hinterland strategy of scent marking is probably a response to the problem of marking a very large area with a limited amount of scent. Border marking may give the earliest warning of trespass, but it involves a single line of defence, which must be maintained intact if overt conflicts are to be avoided (Gorman 1989). As territory size increases it becomes progressively more difficult to do this, and hinterland marking becomes a safer strategy. The larger the territory, moreover, the less intense border marking becomes. In the brown hyaena, border marking is still present, whereas in the spotted hyaena, with its even larger territories, it is absent. The efficiency of the hinterland scent marking strategies is illustrated in the computer simulations.

The two pastes deposited by brown hyaenas and the higher frequency with which they paste suggest a broader function of pasting behaviour in the brown hyaena than the spotted hyaena. The short-acting black paste of the brown hyaena may have an intragroup function. Brown hyaenas often feed on small items with a low rate of renewal (section 2.2.1). It is important, therefore, that the members of a clan know where others have foraged in the recent past, so that they can avoid unproductive areas and lessen the chances of competing for limited food resources. As the black paste loses its odour rapidly, it may contain information on the time that has elapsed since a hyaena passed that way. To pass on such information efficiently requires that pastings be deposited at a high rate and on a regular basis as an individual forages through its territory. Because of the high degree of relatedness between most of the members of a brown hyaena clan (section 7.1.2), the passing on of such information could have evolved through kin selection. In addition, or alternatively, the high frequency of marking within the territory might be for clan members to register their presence in the territory as they only meet each other occasionally.

The more group-orientated spotted hyaenas mark at a lower level and produce less scent. They also tend to concentrate their scent marks at a restricted number of sites of interest spread throughout the territory – a clumped as opposed to a regularly spaced distribution of pastings. This may increase the chances of them being detected compared to the same number thinly spread over such large areas.

In a number of social carnivores dominance is correlated with a high frequency of scent marking, for example, in the wolf (Peters & Mech 1975), and European badger (Kruuk *et al*. 1984). In the brown hyaena there were no differences in the pasting frequencies of different social classes, and little indication of a dominance hierarchy (section 7.2.3). In the spotted hyaena, adult immigrant males marked at the highest rate, but were subordinate to adult females (section 7.3.2). Although adult females, because of their superior numbers in the clan, may leave the same order of magnitude of scent marks in the territory as do the adult males, at the individual level it may be important for adult males to invest more energy in scent marking than do females. Competition amongst immigrant males for a position in a clan must be intense, since it is associated with mating rights (section 7.3.2.4). Immigrant males attempting to join a clan may need to advertise continually their presence to the female members, and one way to do this would be by scent marking. As I have suggested in section 5.2.2.2, vocalizations may also play a role here. Equally, once an immigrant male has managed to become integrated into a clan, he should advertise this fact to potential competitors. One way to do this would be through scent marking.

5.4 Summary

	Brown hyaena	Spotted hyaena
Basic postures similar, but visual communication enhanced by	Pilo-erection	Striking tail movements
Social interactions	Mainly dyadic Simple meeting ceremony	Mainly communal Elaborate meeting ceremony
Vocal communication	Graded system, eight vocalizations Mainly soft and infrequently made Short distance and dyadic only	Graded system, 14 vocalizations Many loud and frequently made Short and long distance, especially the whoop, also group orientated
Chemical communication	Pasting and latrines	Pasting and latrines
	Frequency: 2.64 pastings/km	Frequency: 0.13 pastings/km

| Hinterland marking, marks regularly spaced throughout territory | Hinterland marking, but marks tend to be clumped in distribution |
| Secrete two pastes – short and long acting | Secrete long-acting paste only |

Statistical tests

1. Comparison of the observed versus the expected frequencies with which brown hyaenas partook in neck-biting when in boundary blocks in their territories. The expected frequency was calculated from the relative distance hyaenas travelled in boundary blocks as opposed to internal blocks.
$\chi^2 = 5.67$; d.f. $= 1$; $p < 0.02$; $N = 18$.
2. Comparison of the frequency with which brown hyaenas won neck-biting interactions in boundary blocks versus the frequency with which they won in internal blocks.
Fisher exact probability test: $p = 0.4169$; one-tailed; $N = 18$.
3. Comparison of the frequency with which brown hyaenas, after winning a neck-biting interaction, remained in the vicinity versus the frequency with which they departed immediately.
Fisher exact probability test: $p = 0.0002$; two-tailed; $N = 7$.
4. Comparison of the frequencies with which male and female brown hyaenas presented to the opposite sex versus the frequencies with which mutual or no presenting occurred when individuals of opposite sex met.
Fisher exact probability test: $p = 0.8534$; two-tailed; $N = 24$.
5. Comparison of the observed versus the expected frequencies that brown hyaenas from different age-groups muzzle-wrestled with each other. Expected frequencies were calculated from Table 5.4.
$\chi^2 = 39.00$; d.f. $= 5$; $p < 0.001$; $N = 97$.
6. Comparison of the pasting frequencies of male versus female brown hyaenas.
Mann–Whitney U test: $U = 4$; $n_1 = 3$; $n_2 = 4$; $p = 0.628$; two-tailed.
7. Comparison of the observed distribution of nearest neighbour distances between brown hyaena pastings in the Kwang 2 (1976) territory versus the expected frequency if the pastings were randomly distributed.
$\chi^2 = 419$; d.f. $= 7$; $p < 0.001$; $N = 2978$.
8. Correlation between the number of pastings deposited by brown hyaenas of the Kwang 2 (1976) territory per grid block and the distance they moved in that block.
   (i) *Boundary blocks*. Pearson product–moment correlation coefficient: $r = 0.80$; $p < 0.001$; $N = 27$.
   (ii) *Internal blocks*. Pearson product–moment correlation coefficient: $r = 0.73$; $p < 0.001$; $N = 16$.
9. Comparison of the pasting frequencies of brown hyaenas from the Kwang territory in boundary blocks versus internal blocks.
Mann–Whitney U test: $U = 566$; $z = 4.6$; $n_1 = 16$; $n_2 = 27$; $p < 0.001$; two-tailed.
10. Comparison of the observed distribution of nearest neighbour distances between brown hyaena latrines in the Kwang 2 territory versus the expected frequency if the latrines were randomly distributed.
$\chi^2 = 78.4$; d.f. $= 7$; $p < 0.001$; $N = 70$.

11. Comparison of the observed distribution of brown hyaena latrines around shepherd's trees.
Rayleigh's test: $z = 17.14$; $p < 0.001$; $N = 34$.

12. Comparison of the frequencies with which brown hyaenas in the Kwang territory visited latrines in boundary blocks versus internal blocks.
Mann–Whitney U test: $U = 317.3$; $z = 2.9$; $n_1 = 16$; $n_2 = 27$; $p < 0.002$; two-tailed.

13. Comparison of the observed distribution of nearest neighbour distances between spotted hyaena latrines in the Kousaunt territory versus the expected frequency if the latrines were randomly distributed.
$\chi^2 = 133$; d.f. $= 7$; $p < 0.001$; $N = 117$.

14. Comparison of the observed distribution of nearest neighbour distances between spotted hyaena pastings in the Kousaunt territory versus the expected frequency if the pastings were randomly distributed.
$\chi^2 = 319$; d.f. $= 5$; $p < 0.001$; $n = 1106$.

15. Comparison of the number of latrines in boundary blocks of the spotted hyaena Kousaunt territory versus internal blocks.
Mann–Whitney U test: $U = 43$; $n_1 = 88$; $n_2 = 161$; $p = 0.0012$; one-tailed.

16. Correlation between the number of pastings by spotted hyaenas of the Kousaunt territory per grid block and the distance they moved in that block.
    (i) *Boundary blocks*. Pearson product–moment correlation coefficient: $r = 0.091$; $p > 0.05$; $N = 88$.
    (ii) *Internal blocks*. Pearson product–moment correlation coefficient: $r = 0.221$; $p > 0.05$; $N = 161$.

17. Comparison of the pasting frequencies of the spotted hyaenas from the Kousaunt territory in boundary blocks versus internal blocks.
Mann–Whitney U test: $U = 528$; $n_1 = 88$; $n_2 = 161$; $p > 0.05$.

6 The comparative denning behaviour and development of cubs

Like all large carnivores, hyaenas invest heavily in the development of their young, but the manner in which the two species do so varies somewhat. Both species house their young in dens for extended periods, but the features of these dens, and the roles of the individuals concerned with cub development vary. Unexpectedly, cooperation in raising young is more apparent among brown hyaenas than spotted hyaenas (Mills 1983b).

6.1 Dens

The den is the centre of social activity of a hyaena clan. All the members visit the den, where they meet the cubs as well as other clan members. The prime function of a hyaena den is to provide protection for the cubs during the long periods that the adults are away from it. Although the entrances may be large, they narrow down to small, oval-shaped tunnels, 30–50 cm high, 50–60 cm wide (Watson 1965, Kruuk 1972, Skinner 1976, Henschel *et al.* 1979, Owens & Owens 1979a, Mills 1983b), large enough for cubs, but not for adult hyaenas or other large carnivores, to enter. They are also sufficiently deep to make it unlikely that a large carnivore will dig out the cubs. Small cubs only emerge from the den if their mother, or perhaps another adult from the group, is present (Fig. 6.1), and cubs of all ages immediately run into the den if alarmed.

Table 6.1 compares certain features of brown and spotted hyaena dens. The mean number of cubs per brown hyaena den was 3.0 (Table 6.1), and at most dens there was only one litter. Of 12 reliable cases, three dens (25%) contained two litters, at two of which the breeding females were known to be a mother and daughter. Spotted hyaenas normally den communally (Fig. 6.2); the maximum number of females with cubs at a den was four, and the mean number of cubs per den was 4.4 (Table 6.1). Sometimes, two dens were used simultaneously by members of the same spotted hyaena clan, but this only occurred over periods when dens were being changed. Once, a female with newborn cubs kept them at a den away from the main den for a few weeks, but normally, newborn cubs were found at the communal den.

Spotted hyaenas are more likely to den along a river bed than are brown hyaenas (Table 6.1). Brown hyaenas probably avoid denning in this

Table 6.1   A comparison between certain features of brown hyaena and spotted hyaena dens in the southern Kalahari.

Feature	Brown hyaena		Spotted hyaena	

1. Number of cubs per den

Feature	Brown hyaena	Spotted hyaena
Mean	3.0	4.4
Mode	2	6

Mann-Whitney U test: $U = 375$; $z = -3.26$; $n_1 = 27$; $n_2 = 50$; $p < 0.01$; two-tailed

2. Habitat in which den situated

	Brown hyaena		Spotted hyaena	
	No.	%	No.	%
River bed	0	0.0	3	4.3
Side of river (calcrete)	1	1.9	19	27.5
Side of river (sand)	2	3.8	14	20.3
Pan	1	1.9	1	1.4
Dunes	49	92.5	32	46.4

River vs. dunes $\chi^2 = 27.71$; d.f. = 1; $p < 0.001$

3. Number of den entrances

	No.	%	No.	%
1	42	93.3	15	50.0
2	2	4.4	7	23.3
3	1	2.2	7	23.3
4	0	0.0	1	3.3

One hole vs. more than one hole: $\chi^2 = 20.98$; d.f. = 1; $p < 0.001$

4. Food remains at den (Mills & Mills 1977)

	No.	No.
Mean number of items per den	15.1	1.7

Mann-Whitney U test: $U = 2.5$; $n_1 = 3$; $n_2 = 5$; $p < 0.002$; two-tailed

Percentage composition of bones	%	%
Large herbivores	14	53
Medium-sized herbivores	21	40
Small herbivores	15	7
Small carnivores	38	7
Birds, eggs, tortoises	12	0

Large animals vs. medium-sized and small animals $\chi^2 = 7.61$; d.f. = 1; $p < 0.01$

5. Mean period of occupancy of den

	3.6 months	1.5 months

Mann-Whitney U test: $U = 132$; $z = -4.68$; $n_1 = 47$; $n_2 = 20$; $p < 0.001$; two-tailed

Figure 6.1 A six-week-old brown hyaena cub follows its mother as a second cub emerges from the den on the extreme right.

Figure 6.2 Six spotted hyaena cubs and two adult females at a den in the Nossob river bed.

habitat, to reduce harassment by the other large carnivores (sections 4.4.2, 8.1.2). Spotted hyaena dens also usually contain more entrances (and probably more tunnels) than do those of brown hyaenas, probably because they tend to have more occupants. At the large spotted hyaena dens in East Africa, where up to ten females share a den, the number of entrances may be 12 or more (Kruuk 1972). However, brown hyaenas often have one or two minor dens at distances varying from 5–500 m from the main den, whereas spotted hyaenas do not. Brown hyaena dens usually contain larger (Fig. 6.3) and, on a species basis, different bone accumulations from those of spotted hyaenas (Table 6.1). This is because of differences in the way in which the two species feed their cubs (section 6.2.3).

The average period of den occupancy was short for both species, especially for spotted hyaenas (Table 6.1). This is in contrast to the Kruger National Park where spotted hyaenas usually spend over six months, even several years, at a den (personal observations). Likely reasons for moving dens included disturbance by man (three cases with brown hyaenas – also recorded in brown hyaenas by Goss (1986), and a build-up of fleas at the den (nine cases with spotted hyaenas and two cases with brown hyaenas). Flea infestations may explain the brief occupancy of hyaena dens in the Kalahari, particularly at the larger spotted hyaena ones, as recently abandoned dens had numerous fleas, whereas newly occupied ones were free of them.

Figure 6.3 ' A brown hyaena den littered with food remains brought to the den by adults to feed the cubs.

At least 60% of the 69 dens used by spotted hyaenas were taken over from porcupines. Some of these holes were many years old, with large entrances. In two cases the hyaenas killed a porcupine shortly after moving in, and in five cases spotted hyaenas and porcupines apparently co-habited peacefully. In the best example, up to four porcupines emerged from a den after sunset on three consecutive nights, while adult hyaenas lying at the den ignored them. There was no evidence that brown hyaenas displaced porcupines, but rather occupied abandoned aardvark holes. However, porcupines moved into seven of 14 brown hyaena dens checked after the hyaenas had vacated them.

Most dens were used for only one period by hyaenas. Six (17%) of 36 spotted hyaena dens, and three (12%) of 25 brown hyaena dens were used more than once, and two dens were used by both species. Spotted hyaenas had some favourite dens; one in the Kaspersdraai territory was occupied at least six times over a nine year period, and one in the Kousaunt territory, four times over a three year period.

Brown hyaenas tend to den in one vicinity over several years; the Kwang clan dens were concentrated in only 2 km^2 between 1974 and 1982 (Fig. 6.4), and the Kaspersdraai clan within 5 km^2 in 1972 and 1973. Spotted

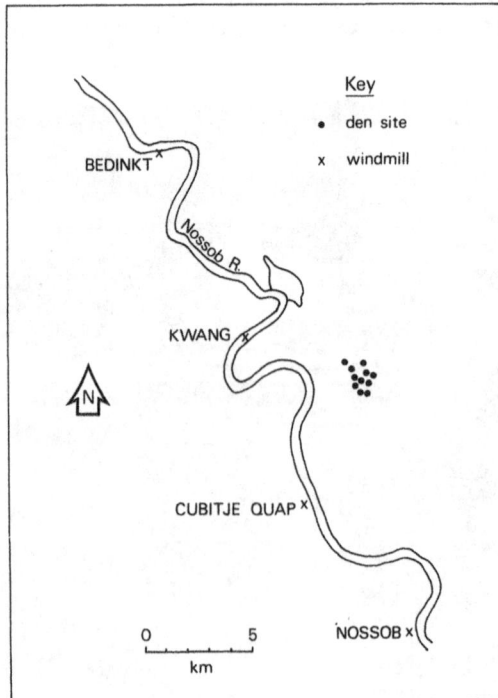

Figure 6.4   The location of dens of brown hyaenas from the Kwang clan, 1974–1980.

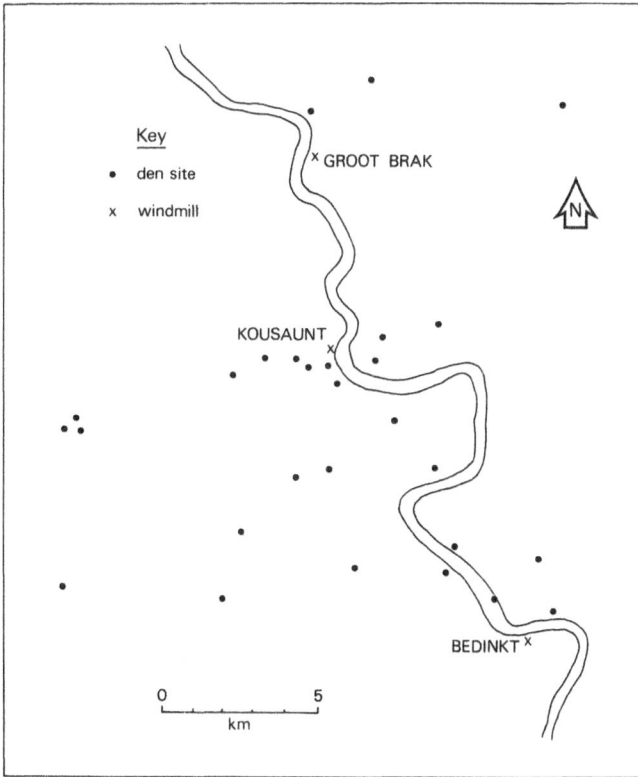

Figure 6.5   The location of dens of spotted hyaenas from the Kousaunt clan, 1979–1984.

hyaenas' dens are less clumped; the Kousaunt clan dens spanned 270 km² between 1979 and 1984 (Fig. 6.5), the Kaspersdraai and Seven Pans clans denned within 45 km² and 90 km² respectively, in 1983.

On four occasions when there were no cubs under nine months of age in a spotted hyaena clan, the animals did not use a den for several weeks, but nonetheless returned daily to particular resting sites. These areas can be equated with rendezvous sites used by wolf packs with large cubs (Mech 1970).

In contrast to my observation, Owens & Owens (1979a) recorded that brown hyaena dens in the central Kalahari were used for several years, that the females utilized one major den located centrally within the clan's territory, and that cubs were born in minor dens and transported by the mother to the communal den. At one den they reported that four females had a total of five cubs ranging in age from four to 20 months, although elsewhere (Owens & Owens 1984) they reported that usually only one female in this clan gave birth to cubs each year.

6.2   Development of cubs

Although physical and social development of cubs does not differ greatly between the two species, their foraging activities do.

6.2.1   Physical development

Brown hyaenas are born with their eyes closed, opening them at eight days, and with their ears bent sharply forward, forming triangles, only becoming erect at four weeks (Schultz 1966, J. Anderson personal communication). The pelage is the same colour as the adults', but the hair is shorter (Fig. 6.1).

Spotted hyaenas are born with their eyes open (Pournelle 1965), and have a brownish-black pelage. Spot development is variable, but usually at about six weeks of age the head starts turning a light grey, and spots start appearing on the neck at about ten weeks (Fig. 6.2). By 16 weeks spots appear on the flanks. The legs are the last part of the body to develop spots, which appear in the second year (Kruuk 1972).

Tooth eruption sequences have not been studied in detail in hyaenas. Brown hyaenas are born without teeth, and milk incisors appear by the fourth week (J. Anderson personal communication). Spotted hyaenas are born with erupting milk incisors and canines, and milk premolars appear in the second month (Kruuk 1972). In both species the permanent canines appear at 10–12 months, and the full permanent dentition is through at about 15 months (Kruuk 1972, Mills 1982c, personal observations).

The average birth mass of spotted hyaenas is quoted by Pournelle (1965) as 1.5 kg, which is approximately 2.5% of the adult mass. The average birth mass of brown hyaenas has been calculated as 693 g (Mills 1982c), which is approximately 1.9% of the adult mass. Brown hyaenas reach full adult size at about 30 months of age (Mills 1982c), as do male spotted hyaenas (Kruuk 1972, personal observations). However, female spotted hyaenas do not reach adult size until they are three years old or more (section 1.4.2).

6.2.2   Social behaviour

Both species show a number of adult behaviour patterns at a very early age. For example, brown hyaenas muzzle-wrestle by six weeks of age, and spotted hyaenas are able to perform the meeting ceremony, with raised hind leg and complete penis or clitoris erection, before they are one month old. These organs, moreover, are relatively larger than in adults (Kruuk 1972). Both species are also able to go through the motions of pasting with complete anal bulging at six weeks of age, although the anal scent gland is as yet inactive. By four months of age brown hyaena cubs are able to secrete the black paste, but the white paste is not secreted until the cubs are over a year old. The anal glands of spotted hyaenas do not secrete until they are over a year old (Kruuk 1972).

Young hyaena cubs are also able to produce some adult vocalizations (Table 5.6), although their pitch is considerably higher than in adults. I have heard spotted hyaena cubs of six weeks old whooping, giggling, and squealing. Other vocalizations in both the spotted hyaena and the brown hyaena, notably the whines and squeals, may be regarded as infantile, but are produced in the same form by adults. Lower-pitched calls such as growls, the low, the hoot-laugh, and the rumble are not heard from cubs, although brown hyaenas as young as four months old may emit the soft growl when muzzle-wrestling.

6.2.3 Nursing and foraging

The denning and suckling periods of hyaenas are longer than in other large carnivores. For the first year to 15 months of their lives, cubs of both species spend most of their time at the den, only being weaned at 12–16 months. In contrast, wolf and wild dog pups only spend 3–5 months at the den, and are weaned at about two months (Mech 1970, Malcolm & Marten 1982), lions are weaned at six months, cheetahs at 2–3 months (Schaller 1972), and even bears, which give birth to highly altricial young, have a denning period of some 4–5 months, and wean their young at 7–8 months (Herero 1972).

The manner in which food is provided to the cubs differs between the brown hyaena and the spotted hyaena. Brown hyaena adults regularly carry food back to the den for the cubs, so that their milk diet is supplemented by meat from approximately 12 weeks of age, whereas spotted hyaenas do not carry substantial amounts of food to the den. On all 12 occasions that they did so the food consisted of skin or bones. Although cubs might chew on these, the nutritional value they obtain is negligible. Accordingly, spotted hyaena cubs rely heavily on a milk diet until they are old enough to accompany adults on foraging expeditions, which only become a regular occurrence when the cubs are about one year old (Fig. 6.8).

When a female brown hyaena (or other adult) arrives at a den containing cubs, it generally puts its head into the hole and utters a very soft growl (Table 5.6). The cubs soon emerge, and, unless the adult is lying down, they follow it with their ears flattened out sideways and lips drawn back, uttering a harsh whine. If the adult is a female, particularly the cubs' mother, she usually allows them to suckle. Normally, the female lies on her side with her uppermost hind leg raised, and with the cubs lying at right angles to her. Rarely, she will stand while they suckle. Suckling is terminated by the female rolling onto her stomach. Female brown hyaenas will suckle cubs other than their own (also reported by Owens & Owens, (1979a)), although they show a clear preference for their own cubs. On only four (27%) of 15 occasions when a cub attempted to suckle from a female that was not its mother, was it able to drink for more than one minute. At two communal dens of the Kwang clan, the daughter (Chinki) allowed her mother's (Normali) (Fig. 4.2) cubs to suckle for 22 and 6 min, and once, the mother allowed one of her daughter's cubs to suckle for 52 min. The following is an example of an interaction between the daughter and the cubs:

> 19 September 1977. 01.12 h. Chinki comes to the den and starts to suckle two of her three cubs. Immediately, one of Normali's larger cubs approaches and also suckles. After 6 min Chinki's third cub comes out of the den, approaches Normali's cub and struggles with it for the teat. Chinki takes no notice. After a short while Chinki's cub wins the teat, but Normali's cub tries to get it back again. Chinki snaps at Normali's cub, which then moves away. It soon returns, and lies down next to the same small cub for a minute, but does not attempt to suckle. After 36 min Chinki terminates the suckling.

Spotted hyaenas call their cubs out of the den with a soft growl or lowing sound (Table 5.6). Prior to suckling, the cubs follow their mother in a similar manner to brown hyaena cubs. However, when suckling, the position adopted by the cubs is different; they lie parallel to the mother with their tails near her head, and sometimes one of the cubs lies between the female's hind legs (Fig. 6.7). Like brown hyaenas, the female occasionally remains standing while suckling, and the bout is terminated by the female rolling over onto her stomach, or by her walking away.

Spotted hyaenas were only observed to suckle their own cubs (see also Kruuk 1972). Cubs showed interest in the teats of other lactating females, and even suckled from them for a few seconds, but they were soon driven away by the female or her cubs, or left voluntarily. Recently, however, allo-suckling was observed for several weeks amongst three adult females of the Kousaunt clan by M. Knight and A. van Jaarsveld (personal communication). The presence of three single cubs of similar ages born to three closely related females may have led to this unusual behaviour.

6.2.3.1 0–3 MONTHS

For the first three months of their lives brown hyaena cubs are visited regularly by their mother, normally at sunrise and sunset. During eight 24 h observation periods at a den with cubs 0–3 months of age, the mother visited the cubs on 14 occasions. At this stage the cubs rarely come out of the den, except when an adult is there. In the Namib Desert, Goss (1986) observed that a female brown hyaena with small cubs spent 82% of the 24 h period at the den.

The presence at a den of spotted hyaena cubs under six weeks of age can often be detected by the behaviour of the mother. She occupies a position right in the den entrance, whereas mothers of larger cubs and other clan members lie further away. The female spends most of the day at the den, unless forced to move away to seek shade. At night she may forage far afield, but quickly returns to the den after feeding. In 19 (68%) of 28 visits I made to a den at sunrise, and in 21 (81%) of 26 sunset visits, a mother of 0–3-month-old cubs was at the den.

The durations of suckling bouts of cubs of various ages are shown in Figure 6.6. Spotted hyaenas generally suckle their cubs for longer periods than do brown hyaenas.

6.2.3.2 4–9 MONTHS

The mothers of four-month-old brown hyaenas visit their cubs once during a 24 h period at night. In 115 24 h observation periods at dens with 4–9-month-old cubs, the mother visited them 86 (75%) times. Brown hyaena females suckle 4–9-month-old cubs for longer periods than they do 0–3-month-old cubs (Fig. 6.6) (1).

At four months the mother and other brown hyaenas begin to bring food to the cubs. Typically, the food is carried to the entrance and dropped as far down the tunnel as the adult can reach. If the cubs are out of the den when an animal arrives with food they run over to it, and the animal drops the food. The cubs then pick it up and carry it into the den. The food is mostly consumed underground and often the cubs take turns being in the den when there is food inside. The remains are eventually brought out of the den.

As the cubs grow older they begin to spend more time out of the den, emerging at sunset even when adults are absent. However, they do not venture far from the den.

The mothers of 4–9-month-old spotted hyaena cubs were present at the den in 31 (67%) of 46 sunrise visits and 44 (75%) of 59 sunset visits. However, as with brown hyaenas, the time spent suckling 4–9-month-old

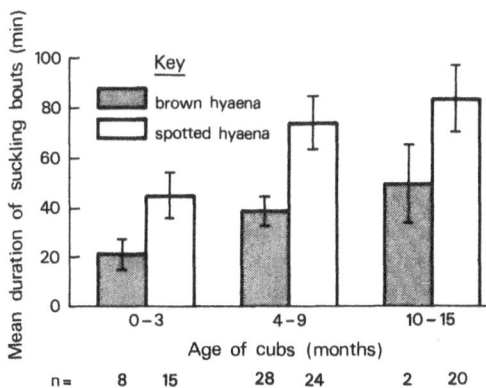

Figure 6.6 Mean ± SE duration of suckling bouts of different aged cubs for brown hyaenas and spotted hyaenas.

cubs was significantly longer than it was for 0–3-month-old cubs (Fig. 6.6) (2). Cubs in this age group will also come out of the den even if no adults are present, and older cubs will sometimes move short distances (1–2 km) away from the den alone. Towards the end of this period spotted hyaena cubs sometimes accompany adults on foraging trips (Fig. 6.8), some of which may be quite extensive. For example, a group of nine adults were accompanied by one ten-month-old, two nine-month-old and two seven-month-old cubs, for 29 km, and a female moved 17 km with her two six-month-old cubs.

Although the cubs may obtain some meat on these expeditions, milk is still the mainstay of their diet. The cubs only feed from the carcass once the adults have fed well. It is probably dangerous for them to feed while the adults are doing so, considering the scramble that characterizes feeding bouts. Additionally, there is the danger of attack by competitors; for example, a seven-month-old cub was killed by a male lion at a carcass (Fig. 2.27).

6.2.3.3   10–15 MONTHS

The visits by brown hyaena mothers to the den become rarer as the cubs grow older. In only 16 (35%) of 46 24 h observation periods did the mother visit the den of 10–15-month-old cubs (3). The contribution of milk to the diet of brown hyaena cubs becomes much less important – in only two (13%) visits did the mother suckle her cubs.

When a brown hyaena cub is approximately 10 months old it spends time away from the den. It may move off a short distance in the morning and spend the day under a bush. The distances moved in the mornings from the den, and particularly the distances moved in the evenings on the return journeys, gradually increase, as the cubs begin to find food. When the cubs are one year old their foraging trips are quite extensive – sometimes they stay away from the den for over 24 h. The mean distance moved during 13 foraging trips by cubs over one year old was 12.4 ± 1.99 km (range 2.7–26.2 km). They normally move alone, occasionally with another cub for a short distance.

In the Kaspersdraai brown hyaena clan, the mother of the cubs died when they were approximately one year old (Fig. 2.7). She was the only animal known to bring food to this den. However, the cubs managed to survive at the den for another six weeks before abandoning it at an earlier stage than usual. Their fate was unknown.

Ten to 15-month-old spotted hyaena cubs are still frequently suckled (Fig. 6.7), for bouts lasting for similar times to those of 4–9-month-old cubs (Fig. 6.6) (4). During this period cubs accompany the adults on foraging trips increasingly often (Fig. 6.8). Thus, although milk still forms a significant part of their diet, meat becomes more important.

Spotted hyaenas are weaned at between 12 to 16 months (Kruuk 1972). A female who suckled her 14-month-old cubs for 90 min, was seen to reject

Figure 6.7 Two ten-month-old spotted hyaena cubs suckling from their mother. One of the cubs is lying between the female's hind legs.

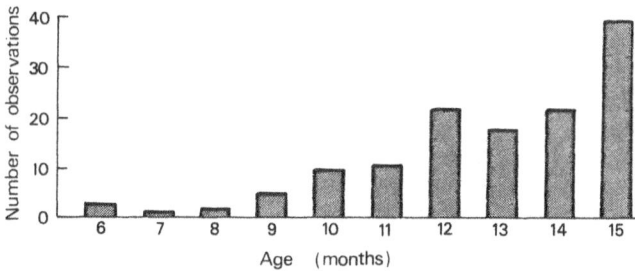

Figure 6.8 The frequency with which spotted hyaena cubs of various ages accompanied adults on foraging trips.

one of them four weeks later, when the cub lay next to her for 10 min begging (whining). I never saw her suckle these cubs again, although on another occasion one begged.

6.3 Sub-adults

At 15 months the denning period is basically over, and from then until 30 months of age animals of both species are regarded as sub-adults. Physical development is more or less complete by 30 months of age (except for spotted hyaena females) and some of them are reproductively active (Ch. 4).

If there is another den in the territory, brown hyaena sub-adults will spend much time there. They will eat food brought to the den, even robbing adults as they arrive. For example, a two-year-old female robbed her mother of a steenbok carcass the adult brought to her seven-month-old cubs. When she was about 50 m away from the den the sub-adult ran over to the adult. She ran past her daughter and dropped the carcass at the den, but her daughter snapped it up and ran off with it.

Three brown hyaenas were first observed to carry food to the den when they were 22, 24 and 27 months old. There is a transition period between taking food that other adults have brought and carrying food to the den, as is illustrated in the following two examples:

10 July 1978. Kwang Den. The mother of four, five-month-old cubs arrives at her den just before sunrise carrying a bat-eared fox. A 16-month-old male (this female's daughter's cub) which is at the den, grabs the fox from the female and runs off with it. After a few minutes he returns with the carcass and the adult female snaps at him. He drops the fox into the den entrance and then moves off.

8 August 1978. Kwang Den. An adult male comes to the den at 00.42 h carrying a springhare. A 22-month-old male, which four nights previously brought food to the den, is there. The sub-adult moves over to the adult male emitting a harsh whine (Table 5.6). The adult drops the springhare and the sub-adult picks it up and carries it off. Half an hour later the sub-adult returns to the den carrying the pelvic girdle and legs of a springbok. He drops the food and goes over to the adult male harsh whining, and presents to him as in the meeting ceremony (section 5.1.3.1). The sub-adult then picks up the food again and carries it over to the den entrance with his ears flattened out sideways and lips drawn back. He drops it in front of a cub which carries it into the den.

Similar observations were made by Goss (1986) in the Namib Desert, and Owens & Owens (1984) in the central Kalahari.

Sub-adult spotted hyaenas rarely forage on their own. Of 72 observations of solitary foraging hyaenas from the Kousaunt clan, only eight (11%) were sub-adults (5). Furthermore, I never encountered a group of foraging hyaenas which comprised solely of sub-adults. It seems, therefore, that sub-adults are dependent on adults for most of their food, particularly hunted prey. This is discussed further in section 7.3.3.2.

As sub-adults, spotted hyaenas may acquire some of the skills needed to hunt large prey. I have no observations of adults creating situations for their young to hunt or kill, as has been recorded in cheetahs and tigers (Schaller 1972), but sub-adults normally stay to the rear of a foraging group, which would be a good place to observe the actions of the adults. Through experience sub-adults may also learn which animals to attack and which not to waste their energy on. Twice, when a foraging group of spotted hyaenas containing adults and sub-adults encountered large potential prey (once an adult gemsbok and once a herd of adult eland), the sub-adults expended considerable energy, and possibly risked injury, in futile attempts to overcome the prey, while the adults paid little attention to them. An example is given below:

> 4 December 1978. 23.45 h. Dunes west of Bedinkt. Five spotted hyaenas (two adults and three sub-adults) detect an adult gemsbok cow 50 m away. The three sub-adults move towards the gemsbok, then stop and look back at the two adults which have lost interest in it. The sub-adults run over to the gemsbok which immediately runs off. They chase it for 1.4 km, then the gemsbok backs up against a small bush and faces them. For the next 14 min the sub-adult hyaenas attack the gemsbok, which alternatively stands and lies on its brisket, swinging its horns at the hyaenas, twice nearly horning one. One hyaena does manage to bite the gemsbok briefly in a back leg, but this is not effective. Eventually the hyaenas move off slowly.

As mentioned, at around 30 months of age both species of hyaena reach sexual maturity and may begin breeding. This is also the time when many individuals leave the comparative shelter of their natal clans to become nomadic, or to begin the difficult process of integration into another clan. The way in which much of this is accomplished is discussed in the next chapter.

6.4 Functional considerations of denning behaviour in the Hyaenidae

There are certain features of the dens and denning behaviour of the Hyaenidae which tend to separate them from other cooperative breeding carnivores which raise their young at dens. These are discussed below.

Hyaenas do not usually leave guards at the den as do other carnivores (Macdonald & Moehlman 1982, Emlen 1984). Although the mothers of small cubs, or other clan members may spend prolonged periods at the den (Goss 1986, Henschel 1986), this is irregular – sometimes there will be several hyaenas at a den and at others there will be none. Perhaps the need to leave guards has been obviated by the nature of hyaena dens – the narrow tunnels are large enough for the cubs, but too small for other large carnivores to enter. Henschel (1986) found a high small-cub mortality at two atypical, large-tunnelled, spotted hyaena dens in the Kruger National Park – only one of seven litters born survived longer than two weeks. However, he believes that this was mainly due to the hyaenas not denning communally, as there were no large cubs to protect the smaller ones. Owens & Owens (1979b) also favour the hypothesis that larger cubs provide protection for brown hyaena cubs, but what protection could a nine-month-old cub give to a three-month-old cub in the face of attack from a lion or adult spotted hyaena, for example? Furthermore, an observation of spotted hyaenas raiding a brown hyaena den (section 8.1.2) does not support this hypothesis, but does show the usefulness of the narrow tunnels. The only member of the Hyaenidae where adults have been reported to perform systematic guarding duties is the aardwolf, where the cubs were vulnerable to predation by black-backed jackals (Richardson 1987).

The denning and suckling periods of hyaenas are longer (section 6.2.3), and the survival rate of cubs higher (Ch. 4) than the other large African social carnivores: lions (Schaller 1972, Rudnai 1973, Eloff 1980, Packer *et al*. 1988), and wild dogs (Frame *et al*. 1979, Malcolm & Marten 1982). Perhaps these phenomena are related. Adult lions, being less mobile than hyaenas, are not able to cover long distances from food to young, and, therefore, are forced to take their young along with them, exposing them to additional dangers. The semi-nomadic existence of wild dogs (Frame & Frame 1981, Reich 1981) probably precludes a long denning period so that the pups have to leave the safety of the den at an early stage.

Within the Hyaenidae there are also differences in denning behaviour. Spotted hyaenas do not usually carry food to their cubs, and den communally, whereas brown hyaenas frequently carry food to their cubs (or, in one unusual case along the Namib Desert coast, to the mother of very small cubs (Goss 1986)), and usually only raise one litter at a den. Even at a communal brown hyaena den, observed over a seven year period in the central Kalahari, usually only one female gave birth to cubs each year (Owens & Owens 1984). Striped hyaenas also carry food to the den, and have only been recorded to den singly (Kruuk 1976).

Because brown and striped hyaenas are solitary foragers, an individual which finds a suitable food item for cubs can usually carry it back to the den. Moreover, at the den competition for the food is between cubs of equal age, whereas at a communal den the larger cubs would be able to monopolize the food.

In spotted hyaenas, where several individuals often feed off a large food item simultaneously, and competition is expressed in the speed of eating (Kruuk 1972 and section 3.7.1.2), it is important for each individual to eat as much as it can, as quickly as possible. Often, therefore, there is nothing left to be taken back to cubs, except for skin and bones, and these the cubs would have difficulty in eating. Even when there is enough food at a carcass, in areas such as the Serengeti and the southern Kalahari, the large distances that spotted hyaenas often move to find food (Kruuk 1972 and section 4.4.1) make it difficult for them to carry food to the den regularly. Furthermore, at the communal dens the larger cubs would get most of the food. Hyaenas do not regurgitate food to their young, perhaps because they have too rapid a digestive system (Kruuk 1972).

Spotted hyaenas do sometimes bring food such as pieces of skin and bones to the den (section 6.2.3). Hill (1980) even observed a spotted hyaena bring an entire domestic sheep to a den in Amboseli National Park. He regarded this as unequivocal evidence of a spotted hyaena providing food for cubs. However, the circumstances surrounding this incident may have been atypical, for example, the hyaena may have been disturbed by people while in the act of killing the sheep, which led to it carrying the food to a familiar place. Henschel (1986) observed a sub-adult female spotted hyaena bring the intact carcass of a civet to her mother which had small cubs. The sub-adult dropped the civet in front of the mother who consumed it. These observations suggest that food carrying is an ancestral behaviour pattern in the Hyaenidae, which is now only fully expressed in brown and striped hyaenas.

Spotted hyaena cubs rely on their mother's milk to a greater extent than do brown hyaena cubs. Spotted hyaenas rarely have more than two cubs in a litter (Smithers 1983 and section 4.2.2), whereas the modal litter size for both brown and striped hyaenas is three (range 1–5) (Smithers 1983, Rieger 1979, Mills 1983b). The comparatively small litter size in the spotted hyaena may have evolved as a result of the heavy dependence of their young on milk. In providing additional nourishment for cubs in the form of meat, brown hyaenas and striped hyaenas can raise larger litters.

Communal suckling occurs in the brown hyaena, but not, for example, as commonly as in lions (Bertram 1976), whereas it is usually absent in the spotted hyaena (Kruuk 1972, Frank 1986b, Henschel 1986, this study). Kin selection has been invoked to explain communal suckling in lions and brown hyaenas (Bertram 1976, Mills 1983b, Owens & Owens 1984). The belief that members of spotted hyaena clans are not closely related has been cited as the reason for the lack of communal suckling in this species (Bertram 1979, Mills 1983b). However, it is now known that there is a high degree of relatedness between the females in a spotted hyaena clan (section 7.1). The fact that milk is virtually the only food that spotted hyaena cubs obtain, may make it imperative that females only suckle cubs which are very closely related to them. These are usually their own.

Several hypotheses have been proposed for the evolutionary advantages of communal denning in the Hyaenidae. In the case of the brown hyaena, Owens & Owens (1979b) suggested that in the semi-arid environment of the central Kalahari, where a mother may have to travel far to feed herself and to collect food for the den, the survival of cubs may be favoured when several females are involved in this role. On the contrary, I submit that communal denning in the brown hyaena can only take place when food is abundant, as group size is correlated with richness of food patches (section 4.5). Under favourable food conditions a female brown hyaena may be able to enhance her inclusive fitness by allowing her grown-up female offspring to remain and breed in her natal territory.

A second hypothesis is that protection of cubs may be enhanced in both species through communal denning (Owens & Owens 1979b, Henschel 1986). This may be so, but, as I have discussed, hyaenas do not employ regular guards.

In spotted hyaenas, where cooperation in food acquisition is sometimes necessary, it would be advantageous to have a central assembly point for the members of the clan – a den would be an ideal location. In both species, a central point where members of a clan could meet and reinforce their presence and status in the group might also be advantageous. Group members might also be able to learn the whereabouts of food sources from others at a communal den, i.e. the information-centre hypothesis. This has been cited as a causative function of communal nesting or roosting in birds (Ward & Zahavi 1973, Krebs 1974, De Groot 1980). A possible example of this in the present study is the observation of three spotted hyaenas tracking an obviously well-fed individual from a den to a carcass (section 3.3.1).

It seems that several small benefits may accrue to communal-denning hyaenas, provided that there is sufficient food in the territory to support more than one breeding female. These obviously outweigh any possible disadvantages which communal denning may entail, such as the promotion of a build-up of fleas at the den, and the competition between cubs for food.

6.5 Summary

	Brown hyaena	Spotted hyaena
Dens	Only large enough for cubs No systematic guarding Almost exclusively in dunes Often littered with food remains	Only large enough for cubs No systematic guarding In dunes and river habitat Food remains few
Solitary/communal	Usually solitary, occasionally communal	Communal
Denning period	15 months	15 months
Suckling period	12 months, rare after 9 months	12–16 months, still regularly at 9–12 months
Litter size	Range: 1–4 Mode: 3	Range 1–2
Method of feeding cubs	Suckling and carrying food	Suckling only
Suckle each other's cubs	Occasionally	No, or very rarely
Cubs begin foraging	Alone	With adults
Sub-adults	Begin to bring food to den	Learn hunting skills?

Statistical tests

1. Comparison of suckling bout times of brown hyaena cubs 0–3 months old versus suckling bout times of cubs 4–9 months old.
    Mann–Whitney U test: $U = 53.5$; $z = -2.14$; $n_1 = 8$; $n_2 = 28$; $p = 0.032$; one-tailed.
2. Comparison of suckling bout times of spotted hyaena cubs 0–3 months old versus suckling bout times of cubs 4–9 months old.
    Mann–Whitney U test: $U = 110$; $z = -1.87$; $n_1 = 15$; $n_2 = 24$; $p = 0.0307$; one-tailed.
3. Comparison of the frequency of visits per 24 h period by brown hyaena females to cubs 4–9 months old versus the frequency of visits to cubs 10–15 months old.
    $\chi^2 = 20.95$; d.f. $= 1$; $p < 0.001$; $N = 161$.
4. Comparison of suckling bout times of spotted hyaena cubs 4–9 months old versus suckling bout times of cubs 10–15 months old.
    Mann–Whitney U test: $U = 214$; $z = -0.61$; $n_1 = 20$; $n_2 = 24$; $p > 0.05$.
5. Comparison of the observed frequency with which sub-adult spotted hyaenas from the Kousaunt clan foraged on their own versus the expected frequency. The expected frequency was calculated from the number of observations of solitary foraging hyaenas in the clan and the proportion in which sub-adults were represented in the clan.
    $\chi^2 = 14.38$; d.f. $= 1$; $p < 0.001$; $N = 72$.

7 The individual in hyaena society

In this chapter I investigate how individual hyaenas behave to maxir
the number of their genes passed on through the population, and br:
consider the evolution of the social organizations of the two spe⁤
showing the importance of feeding habits in moulding them. In orde
interpret many of the interactions between individuals, it is necessar
know how they are genetically related to one another, as an animal's g⁤
are passed on both via its own young and its relatives' young, and selec
for genes which cause relatives to be favoured would be expected (Bert
1976). It is also relevant to discuss the mating system and ma⁤
behaviour, including reproductive strategies of males and females, ⁤
parental care, dominance and dominant relationships, and interact
between individuals, both within and outside the clan.

7.1 Degrees of relatedness between clan members

The assumptions made and the calculated degrees of relatednes⁤
between the members of clans in both species are given in Appendi
The average degree of relatedness between 91 pairs of brown hyaenas ⁤
the Kwang clan was 0.26. Owens & Owens (1984) observed a br
hyaena clan over a seven year period which also contained mainly cl⁤
related animals, although no average figure was calculated. The figure
231 pairs of spotted hyaenas from the Kousaunt clan and 32 pairs fron
Kaspersdraai clan were 0.29 and 0.33 respectively. Bertram (1979) n⁤
tained that the average degree of relatedness between the membe⁤
spotted hyaena clans in the Serengeti was less than 0.03, but he had no
to substantiate this.

By way of comparison with other social carnivores, Bertram (1⁤
calculated the average degree of relatedness amongst females in a ty⁤
lion pride as 0.15, and amongst cubs and males as 0.22. However, P⁤
& Pusey (1982) showed that the male lions in a pride are not necess⁤
related, which will affect Bertram's calculations downwards, particu⁤
those for males. Using data from Frame & Frame (1976), Bertram (⁤
calculated that adult male wild dogs in a pack are related on averag
0.38 (the same as my calculations for spotted hyaena females – Appe
E), and the females by 0.5. The average coefficient of relatedness w⁤
red fox social units was also calculated as 0.38 (Macdonald 1980a). He⁤
and cubs in black-backed jackal and golden jackal social units, an⁤

females in Arctic fox groups, share a coefficient of relatedness of 0.5 (Hersteinsson & Macdonald 1982, Moehlman 1983). Both species of hyaenas, therefore, live in groups comprising relatively closely related individuals.

7.2 Brown hyaena society

Although they are solitary foragers, brown hyaenas have an advanced social system with intricate relationships between individuals (Mills 1982b, 1983a, b). In this section I describe some of these relationships, particularly those that might be important for enhancing an individual's inclusive fitness.

7.2.1 The mating system and mating behaviour

Observations of brown hyaenas mating are few, but they suggest an unusual mating system with regard to the males. Most male brown hyaenas eventually leave their natal clans, although some do not do so until they are well into adulthood. The number of adult males in a clan is variable, some clans having none (section 4.1). On leaving their clan the males either become nomads, or join another clan, most brown hyaenas in the southern Kalahari apparently choosing the nomadic option (section 4.3.1). Which individuals are responsible for mating?

Strange males engaged in sexual behaviour with group-living females far more often than did group-living males (Table 7.1) (1). Five of the 11 strange males observed in sexual encounters with group-living females could be individually recognized; three were never seen again, and two were seen twice each in the same area over periods of 12 and 33 days respectively, before disappearing. One of these latter two was seen again once in the same area 27 months later. All observed mating, therefore, was carried out by nomadic males, which visited the clans sporadically. The two females from the Kwang clan mated with at least four different males over a two year period.

As in lions (Hanby & Bygott 1987), potential mating partners who had known each other since birth, i.e. natal males and females, showed no sexual interest in each other, thus precluding the danger of inbreeding (Partridge & Halliday 1984, Land 1985). However, immigrant males would have been expected to show sexual interest in the females. My failure to record sexual behaviour in immigrant males may be because the intensively studied Kwang clan apparently did not have any immigrant males until mid-1980 (Fig. 4.2). This was towards the end of the brown hyaena study, and few observations were made of the new immigrant male (Thirdman). The status of the old male, Hop-a-long (see Fig. 4.2), who died in September 1978, was uncertain. He was observed longer than any other adult

Table 7.1   The frequency with which group-living and strange brown hyaena males were observed
in sexual encounters with group-living females, and the number of interactions observed
between these animals where no sexual behaviour occurred.

	Sexual behaviour			Non-sexual behaviour
	Mounting	Foreplay	Sniffing female's vulva and/or urine	
Group living male	0	0	3	48
Strange male	6	2	3	13

male, and, although he carried food to the den, the only sexual behaviour
he performed was flehmen on one occasion. During the time that he was in
the clan, moreover, four of the six matings seen took place. Assuming that
immigrant males would seek matings it seems more likely that he was a
natal male. Owens & Owens (1984) maintain that immigrant males are the
primary mating males in brown hyaenas in the central Kalahari, but that
they also observed a nomadic male mating.

Some female brown hyaenas were recruited into their natal clan,
probably for life. These females were able to breed, albeit at different
rates. The oldest female (Normali) in the Kwang clan produced cubs at
intervals of 18, 12, 17 and 23 months, whereas her daughter (Chinki) only
produced two litters 41 months apart. In the Seven Pans clan, the older of
the two females raised a litter of cubs, when for 18 months the younger
female failed to reproduce. The mechanism causing females to breed at
different rates is unknown, but does not seem to be related to overt
dominance (see section 7.2.3). The fate of females which left their natal
clan is also unknown. In their study Owens & Owens (1978) reported that
one female joined a clan. However, in a later paper (Owens & Owens
1984) they stated that all females in this clan were related to each other.

The oestrous period of a female brown hyaena lasts several days. Yost
(1980) recorded mating behaviour over at least 15 days in captivity and W.
Ferguson (personal communication) saw a female in the southern Kalahari
mating nine days apart. In between times he saw the female on her own. In
the Kwang clan's territory there was a higher ratio of strange to group-
living males sighted in those months during which mating was observed,
than in months when no mating was observed (2). Males probably learn of
females' oestrus condition by sniffing their urine or vulva and exhibiting
flehmen (Walther 1984).

Twice, a male and female brown hyaena indulged in what appeared to be
courtship behaviour, characterized by some aggression on the part of both
animals and repeated mutual approaching and retreating. One interaction
lasted over 4 h and the other at least 55 min. An increase in agonistic

encounters has been suggested as a precursor to mating behaviour in captive brown hyaenas (Yost 1980). This may be a way in which two strangers overcome their antagonism before mating.

The following is an account of a mating bout:

16 November 1979. 05.15 h. One of the Kwang females is found near the den moving quickly after an unmarked male. She soon catches up with him, but as he turns round to face her she retreats a few metres. The male moves over to her and immediately mounts her. His mouth is open slightly, so that his teeth are bared, his ears are flattened out sideways and he is growling and uttering a staccato-sounding, soft yell (Table 5.6). He bites softly at her back. After 15 s he slips back slightly and clasps the female around her stomach with his forefeet, resting his chin on her shoulder. The female stands with her head down. At first the female's tail is elevated, but after the male has slipped back it is dropped. At this stage intromission is impossible. Three minutes after mounting, the male climbs off the female, the female lifts her tail and the male sniffs and licks her vulva.

The two hyaenas move off, walking parallel with each other about 30 m apart. After a short distance the female stops. The male comes running over to her and mounts her again, but only stays on for 30 s. After dismounting, the two hyaenas stand for 30 s, with their hind regions close together, their heads down and facing away from each other.

The female moves off and is followed by the male. She lies down and the male mounts her. As he does so she stands up and walks forward, with the male clasping her around the stomach, but he soon falls off. The female lies down again and the male lies down behind her, with his head facing her hind region. The two hyaenas groom themselves, then lie still for 10 min, before the female moves off followed by the male.

I follow them until 06.55 h and the male mounts six more times, but never stays on longer than 45 s. During the interaction the male pastes 11 times, but the female never does so.

In Table 7.2 some details of six mating bouts are recorded. I was present at the beginning of three of the bouts – in two the male obviously located the female first, but in the third (recounted above), the female found the male. It was impossible to tell if intromission occurred at each mounting, but in at least 12 (34%) cases it did not, seven times because the male was badly orientated, and five times because the female did not stand still. Pelvic thrusting was never observed. In between mountings the hyaenas often groomed themselves, particularly their genitalia, before moving on, with the female dictating the direction of travel. In two of the six mating bouts, a second strange male was seen briefly in the vicinity of the mating couple. These observations are similar to those of brown hyaenas mating in captivity (Yost 1980, Eaton 1981).

Table 7.2   Details of six mating bouts between brown hyaenas.

Duration of observation (min)	Mounting frequency (mounts/min)	Mean interval (min) between mountings	Duration(s) of mounting			Pasting frequency (pastes/min)	
			Mean	Minimum	Maximum	Male	Female
20	0.15	6.7	-	-	-	0.00	0.00
5	0.40	2.5	9	2	15	0.00	0.00
60	0.12	8.6	64	10	180	0.05	0.00
20	0.20	5.0	11	5	15	0.00	0.00
90	0.10	10.0	87	15	180	0.12	0.00
67	0.15	6.7	37	30	60	0.18	0.03
Mean	0.13	6.6	42	-	-	0.10	0.01

Not all matings are successful; one female mated twice, seven months apart, without producing cubs, and in captivity unsuccessful matings have been recorded at two- and four-month intervals (Lang 1958, Shoemaker 1978, Eaton 1981).

7.2.2   Feeding of cubs by carrying food to the den

Who carries food to the den, how frequently, and under what conditions? The food brought to cubs consists mainly of small or medium-sized animal remains (Table 6.1) which still have some meat on them (Fig. 7.1). In 41 cases where a brown hyaena of food-carrying age found food which was judged to be suitable to be taken to a den, it did so in 27 (66%) of the cases; in 11 (41%) without eating anything first, and in the remaining 16 (59%) after eating part of the carcass first. There was, therefore, a tendency for the hyaenas to take at least some of this food to the den, and they obtained less from the carcass than they would have if they had eaten it all themselves. The mean distance that the food was carried was 6.4 km ± 0.64 (n = 22) – the most spectacular feat being when a brown hyaena carried the remains of a domestic cow calf, which weighed approximately 7.5 kg, a distance of 15 km. Food carrying, therefore, may entail a significant energy cost to the carriers.

Of the cases cited above, males were involved 18 (44%) times and females were involved 23 (56%) times (Table 7.3). The males carried food back to the den 11 times (Fig. 7.2) and females 16 times. Males, therefore, were just as inclined to carry food to the den as were females (3).

Figure 7.1   A brown hyaena female carries the head and shoulder of a springbok to her den.

Table 7.3   Number of times known brown hyaenas carried food to cubs.

Animal	Sex	Number of suitable carcasses found	Number of carcasses carried to the den
Normali	Female	20	14
Chinki	Female	3	2
Hop-a-Long	Male	4	1
Charlie	Male	7	3
Shimi*	Male	5	5
Thunberg	Male	1	1
Thirdman	Male	1	1
Total		41	27

* Four cases in Kwang clan as a natal male, one case in Rooikop clan as an immigrant male

Figure 7.2 An adult male brown hyaena brings an African wild cat carcass to a den containing his mother's and sister's cubs.

Observations at dens also revealed that males bring food to the den; at the Cubitje Quap den the same male was seen to bring food to the den three times, and at the Rooikop den a male was seen to bring food once. In at least two cases immigrant males (Shimi and Thirdman) brought food to a den. These observations are in contrast to those of Owens & Owens (1984), who recorded that females brought food to the den significantly more often than males, and who never saw an immigrant male provisioning cubs.

Owens & Owens (1984) also recorded that males more distantly related than half-sibs did not provision cubs. The relationship of the male Hop-a-long to the cubs he provisioned is uncertain, although it was probably less than that of half-sib (Appendix E). The only male in the Kwang clan who was known to be more distantly related from cubs than the equivalent of half-sib ($r = 0.25$), and who was old enough to provision food during the study, was Shimi (Fig. 4.2). He was related to Chinki's cubs by $r = 0.125$, and to Normali's by $r = 0.25$ (Appendix E). Shimi was observed to provision cubs in the Kwang clan on four occasions. In the first, as he approached a den at which both Chinki and Normali had cubs, one of Chinki's cubs came running over to him. He dropped the food and the cub took it. In two other instances he approached the same den with food and was again met by Chinki's cubs, but tried to evade them, clearly not wanting to give them the food. On both occasions he was robbed. In the final observation, Shimi by-passed the den in which Chinki had cubs and took the food to another den 1.1 km away in which Normali had small cubs. Thus, although he did once provision his half-sister's (Chinki's) cubs, he apparently showed a preference for his mother's (Normali's) cubs.

The presence of helpers at dens has been recorded in several other carnivores, particularly canids and viverrids, and has been reviewed by Macdonald & Moehlman (1982), and Emlen (1984). An important question here is does the presence of helpers lead to an increase in survival of young? Moehlman (1979, 1981) found that pup survival correlated significantly with the number of helpers in black-backed jackals. In coyotes, Bekoff & Wells (1980), and in wild dogs, Malcolm & Marten (1982), found a positive, but non-significant correlation between pup survival and the number of helpers. However, in none of these studies was it established whether the presence of helpers had a direct effect on pup survival, or whether food-rich territories could support more helpers and also more offspring. Harrington et al. (1983) found no evidence for the presence of non-breeding wolves increasing pup survival, and suggested that prey availability influences the ability or willingness of pack members to provide food. Moehlman (1983) also found no correlation between the number of helpers and pup survival in golden jackals, suggesting that death due to exposure and illness negated the apparent benefit of helpers.

It is difficult to assess the effect that brown hyaena helpers had on cub survival. The number of cubs at a den largely depends on the number of females breeding at the time, and the lack of seasonality means that the litters are often of different ages. Also, cub mortality is generally low (section 4.1.5). My limited data show no correlation between helpers and cub mortality (Fig. 7.3).

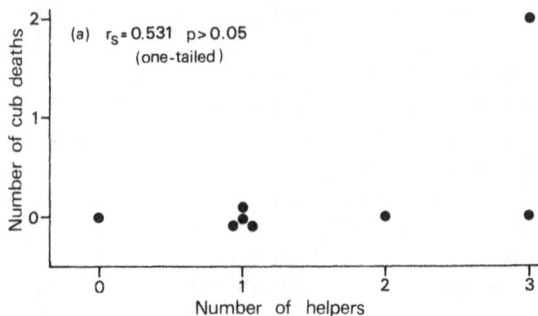

Figure 7.3   The relationship between the number of brown hyaena helpers at a den and the number of cubs that died under 15 months of age in litters that were found for the first time before the cubs were four months old.

7.2.3   Dominance and related behaviour

'To dominate is to possess priority of access to the necessities of life and reproduction' (Wilson 1975: 287). The question of dominance is important in understanding the social organization of group-living animals, many species being arranged in stable dominance hierarchies. However, these hierarchies are frequently complex and difficult to unravel. Owens & Owens (1978, 1979b) reported that a dominance hierarchy, maintained mainly through neck-biting, occurred in brown hyaenas in the central Kalahari, although the evidence they presented to support this conclusion was incomplete. Owens & Owens (1978) said that the hierarchy was not sexually delineated and that some members of each sex dominated members of the other. However, in a later paper (Owens & Owens 1984), they said that the females maintained their own dominance hierarchy. Goss (1986) found a non-linear hierachy, largely based on age, in brown hyaenas on the Namib Desert coast. This hierarchy was not maintained through neck-biting, and interactions between dominant and submissive animals never escalated beyond muzzle-wrestling.

What evidence is there for a dominance hierarchy amongst brown hyaenas in the southern Kalahari? There, neck-biting is predominantly an extragroup behaviour (section 5.1.2) associated with territorial defence rather than maintenance of dominance within a clan. Dominance can be measured in potentially competitive situations such as at food (Rowell 1974, Wilson 1975, Kaufman 1983), the dominant individual or group having access to the resource (Richards 1974). Although brown hyaenas mostly feed alone, several will sometimes assemble at a large carcass. The following example illustrates the behaviour of brown hyaenas around such a carcass:

30 July 1978. Cubitje Quap. 05.10 h. An adult male brown hyaena from the Kwang clan comes to the remains of a wildebeest. A sub-adult male from the clan is feeding, but moves away when the adult arrives. Twenty minutes later the sub-adult returns and starts feeding on some bones 15 m away from the adult. After a few minutes the adult moves away, and the sub-adult returns to the carcass and starts feeding again. The adult sniffs around, picking up a few scraps, and then lies down 10 m from the carcass. After 13 min he stands up, comes to the carcass and starts feeding with the sub-adult. The sub-adult harsh whines (Table 5.6), and the older hyaena attempts to push him away with his hind region. They pull on the carcass and the head becomes detached from the vertebrae. The sub-adult carries it off 10 m and starts eating it, the adult feeding on the vertebrae. After 4 min the adult moves towards the sub-adult. The younger animal raises his hair and tail and hoot-laughs (Table 5.6), at which the adult retreats and goes back to the vertebrae. After 9 min the sub-adult moves rapidly towards the adult with his body held low, hair slightly raised and ears flattened out sideways, uttering a harsh whine (the food acquisition interaction of Owens & Owens (1978)), and grabs at the vertebrae. The two hyaenas pull in opposite directions at the carcass, and the adult soon breaks off a small piece, and moves off with it. Five minutes later he returns, but after a minute or two the sub-adult moves off, and lies down close by.

The most striking feature of brown hyaenas feeding at a carcass is that usually only one hyaena feeds at a time. Typically, a hyaena will break a piece off a carcass and carry it away for consumption or storage. Any other brown hyaenas in the vicinity will wait for the carcass to be vacated, before coming to feed (Fig. 3.35). If two or more hyaenas do feed from a carcass simultaneously, low-intensity aggression will be displayed in the form of pilo-erection, growling and hoot-laughing (Fig. 7.4). In dyadic interactions over food between animals of different ages and sexes no social category of animal had a clear priority (Table 7.4). Neither was there any evidence that certain individuals had priority at carcasses, although the data were too few to analyse this statistically.

Muzzle-wrestling is an important intra-group behaviour pattern amongst brown hyaenas. The functions of this behaviour have been discussed in

Figure 7.4 A sub-adult female (left) and adult female (right) brown hyaena share a springbok carcass.

Table 7.4   Analyses of the outcomes of dyadic interactions at food between different sex and age categories of brown hyaenas of the Kwang clan. The analysis considers which of the two hyaenas fed during an interaction. If a hyaena was feeding and another approached, but then moved away, the possessor received a score. If they fed together, both received a score. If the original possessor then moved away, the new possessor received another score, etc.

Age and/or sex of possessor of carcass	Age and/or sex of animal arriving at carcass	Possessor's score	Arriver's score	χ^2 d.f. = 1
Adult or sub-adult	Cub	7	2	-0.4154*
Cub	Adult or sub-adult	5	2	
Adult	Sub-adult	10	10	1.56; $p > 0.05$
Sub-adult	Adult	16	9	
Adult or sub-adult male	Adult or sub-adult female	10	12	2.95; $p > 0.05$
Adult or sub-adult female	Adult or sub-adult male	19	10	
Adult or sub-adult male	Adult or sub-adult male	16	13	0.31; $p > 0.05$
Adult or sub-adult female	Adult or sub-adult female	8	8	0.00; $p > 0.05$
Total		88	69	2.29; $p > 0.05$

*Fisher exact probability test

section 5.1.3.2, where it is considered *inter alia* to play a role in sub-adult recruitment. In three of seven bouts between an adult female and a sub-adult female, the interaction developed into an aggressive encounter. Two cases involved the eldest daughter (Chinki) of the Kwang clan's matriarch (Normali), and two younger half-sisters of Chinki's. The third case involved the younger of the two Seven Pans females and a young female of unknown relationship. In all three cases the adult clearly dominated the sub-adult. The fact that it was the second oldest female in both the Kwang and Seven Pans clans that behaved aggressively to younger females may be related to eviction. There may have been an incentive to evict younger females which might compete with them for breeding opportunities. In the two cases involving the Kwang clan animals at least, the younger females disappeared from their natal clan. One example of these interactions is given below:

> 6 August 1977. Kwang den, 06.15 h. A sub-adult female (Sanie) begins muzzle-wrestling with two cubs. After 5 min her older half-sister, the adult female Chinki, moves over to the active animals. She grabs the sub-adult on the side of the neck and starts pulling her around while making a panting, growl-like sound. The sub-adult whines, 'grins' and lays her ears back, as the adult pulls her down into the den. Loud harsh whines and yells (Table 5.6) emit from the den for a few minutes, then there is silence. Seven minutes later Sanie comes running out of the den followed by Chinki. The sub-adult runs over to a large bush and the chase continues around it, Sanie with her ears flattened out sideways, Chinki with her ears cocked and slight pilo-erection on the back. After 6 min the adult gives up the chase and moves away.

Apart from these few interactions, there is no evidence that certain individuals dominated others in southern Kalahari brown hyaena clans.

7.2.4 Interactions between animals not belonging to the same clan

The origins of most individuals that interacted with known clan members were uncertain. However, there were indications that clan-living hyaenas distinguished their neighbours from itinerants.

In ten encounters between two animals from neighbouring clans, the outcome varied depending on whether or not they were of the same sex (Table 7.5). Animals of the same sex always acted agonistically towards each other (six observations), whereas animals of opposite sex were never agonistic towards each other (four observations) (4). Where the origin of the intruder was unknown, females always behaved agonistically towards females, males behaved agonistically towards some males, and animals of opposite sex were normally not agonistic towards each other (Table 7.5). Twice, two males known to be residents from different clans met; once

Table 7.5 Nature of interactions between brown hyaenas from different clans and between clan-living individuals and others of unknown origin.

		Agonistic interactions				Non-agonistic interactions		
		Neck-biting	Other forms of aggression	Avoid each other	Total	Ignore each other	Meeting	Total
Both participants from different, but known clans	Male vs. male	1	0	1	2	0	0	0
	Female vs. female	2	1	1	4	0	0	0
	Female vs. male	0	0	0	0	3	1	4
Origin of one of the contestants unknown	Male vs. male	0	1	1	2	4	1	5
	Female vs. female	1	2	0	3	0	0	0
	Female vs. male	1	4	0	5	14	3	17

they neck-bit, and once one of them avoided contact with the other. If neighbouring male residents were generally agonistic towards each other, perhaps the unidentified male intruders involved in non-agonistic inter-actions with residents were nomads. Certainly, in all the cases where I believed that the strange male was a nomad (see section 4.3.1), this was so.

The following is a description of an interaction between members of the Kwang clan and two suspected nomadic males:

23 January 1976. Two adult males (Charlie and Hop-a-long), and the two adult females (Normali and Chinki) of the Kwang clan, are at a gemsbok carcass 5 km south of Bedinkt. At 02.41 h an unknown male arrives. Charlie, who is the only hyaena eating, moves over to the stranger, whining (Table 5.6), and presents to him. The strange male ignores him and runs to the carcass.

The five hyaenas spend the rest of the night at the carcass, feeding, or lying down close by, with the strange male behaving, and being treated, as if he was a member of the clan. The only aggression occurs when the male Hop-a-long clashes with the stranger for 1–2 s, with the Kwang male adopting a low body position.

At 04.37 h a second, smaller, unknown hyaena approaches the carcass. The first unmarked hyaena runs over to the newcomer and grabs it by the side of the neck, with the newcomer yelling (Table 5.6). At this Charlie runs over and joins the attack. After 1 min the two leave the attacked hyaena, which moves away bleeding slightly from one ear.

The next evening I caught and marked the first unmarked male (which was an age-class 4 animal – see section 1.3), but never saw him again. The animal that was attacked by the nomadic male may have been another nomadic male, although its sex was not discerned. Except for the above account there was no evidence that nomadic males are aggressive towards each other. On three occasions two came to a carcass and behaved towards each other as if they were from the same clan, even though in two cases an adult female was present (Fig. 7.5).

It is noteworthy how rarely brown hyaenas from neighbouring clans met during the approximately 1500 h for which adult brown hyaenas away from their dens were observed (Table 7.5). Indeed, I was more likely to encounter a strange brown hyaena in the Kwang territory when not following a hyaena from the Kwang clan than when I was, and vice versa for meeting up with other Kwang clan members (Table 7.6). It seems, therefore, that brown hyaenas avoid hyaenas from neighbouring clans. As discussed in section 5.3.4 this may be accomplished through scent marking. Even when they do meet, encounters rarely escalate to physical combat – of the 14 agonistic encounters detailed in Table 7.5, only five (36%) resulted in neck-biting.

Figure 7.5 Two nomadic male brown hyaenas (left and centre) at a springbok carcass with a group-living adult female.

Table 7.6   The number of brown hyaenas from the Kwang clan and the number of strangers seen at night in the Kwang clan's territory when following Kwang members, or when alone.

	Following a hyaena $n = 462$km	Alone $n = 2491$km	χ^2
Number of Kwang members seen	49	35	7.55; d.f.=1; $p<0.01$
Number of strangers	1	10	

7.2.5   Reproductive strategies and success

Taking the factors discussed in this section into consideration, what are the ways in which individual brown hyaenas can maximize the number of their genes passed on to future generations?

7.2.5.1   MALES

There are three reproductive strategies open to male brown hyaenas:
1. Remain with natal clan. Sub-adults may delay their departure from the clan for several years, or even indefinitely if the resources allow it. Although they do not have any mating opportunities, by helping to raise

related young they may be able to enhance their inclusive fitness through kin selection, although the data regarding helpers and cub survival do not support this contention. Alternatively, helping to feed cubs may relieve the investment of related females, thus allowing them to breed more frequently. They may also be able to monitor conditions in neighbouring territories, transferring when conditions are favourable, as the male, Shimi, may have done when he transferred from the Kwang to the neighbouring Rooikop clan.

2. Become nomadic. This appears to be the most common strategy in the southern Kalahari, and at first sight would seem the best; it involves no investment in offspring, since the females and group-living males attend to their welfare, while the nomads travel on to the next receptive female. However, the lifetime reproductive success of nomads may be low; there are long and erratic intervals between births, not all copulations are successful, and it is possible for more than one male to mate with a female during an oestrous period. In addition, nomadic males forego the possible foraging advantage of having a well-known territory.

3. Immigrate. Immigrant males may have access to females in their group. Even though their chances of locating an oestrous female in their group must be better than those of nomadic males, the expansive territories and widely scattered members make it possible for immigrant males to be cuckolded by nomadic males. In addition, there are long and erratic intervals between births. Immigrant males may do better in areas of higher brown hyaena density such as the central Kalahari (Owens & Owens 1984).

7.2.5.2 FEMALES

Female brown hyaenas have two basic reproductive options:

1. Remain with natal clan. If resources allow it a female may remain with her natal clan and breed. The closely related females cooperate in raising cubs by sharing dens, occasionally suckling each other's cubs and carrying food to the den, and also share in the defence of the territory. Although the younger females may not be able to breed as often as the matriarch, they may be able to inherit her position when the matriarch dies.

2. Leave to establish own territory. The chances of finding a vacated territory in an undisturbed population with a low rate of turnover must be slight. However, success ensures high reproductive rewards. Nomadic females will not be able to raise cubs, although they may eventually find a vacated territory, or be able to return to their natal territory (see section 4.1.3).

7.3 Spotted hyaena society

Spotted hyaenas have a complex society with individuals playing different roles. When analysing their society, therefore, it is important to consider the individual and how its relationships with others may affect its fitness.

7.3.1   The mating system and mating behaviour

Southern Kalahari spotted hyaena clans normally have only one immigrant male (section 4.2.1), but some have none. The evidence, albeit scanty, suggests that the immigrant males do most of the mating, being involved in the three bouts of mating behaviour I observed. Nomadic males may also occasionally mate, as both the Seven Pans females gave birth when there were no immigrant males in the clan (Fig. 4.5). Henschel & Skinner (1987) differentiated between central immigrant males – immigrant males which became residents, and peripheral immigrant males which associated with the clan temporarily. Central immigrant males were involved in the three mating attempts they witnessed. The large clan studied by Frank (1986b) had a number of immigrant males arranged in a dominance hierarchy, with the dominant (alpha) male probably doing most of the mating.

Female spotted hyaenas are often recruited into their natal clan for life, successful immigration to another clan being rare (section 4.2.4). All adult females in a clan may breed, even in large clans (Frank 1986b), although their reproductive success varies (section 7.3.3.1).

Like brown hyaenas, spotted hyaenas are polyoestrous, with each cycle lasting about 14 days (Lindeque 1981), but not all matings are successful. In the three bouts of mating behaviour I observed, the immigrant male, Hans, was involved twice, and the peripheral immigrant male, Nicholas, once. In only one case, involving Hans, did the female subsequently give birth. Two of these observations are recounted below:

18 July 1981. Kousaunt. 23.45 h. An adult female, an immigrant male (Hans), and two sub-adult males are moving together. The female urinates and the three males sniff at the urine and exhibit flehmen. Later, the female urinates again and the adult male exhibits flehmen. The female and the two sub-adult males lie down, and the adult male approaches her with his penis erected. He stops 3 m behind her and licks his forelegs several times. Then he moves forward a few steps, stops again, turns around, backs off, and scrapes the ground with his fore-paws. He stands behind the female looking away from her. After a few minutes he approaches her again, backs away, pastes, and scrapes the ground. Several more times he approaches the female then retreats. Eventually he comes right up to her, attempts to mount while she is lying, then jumps away at the last moment. The female takes little notice. He comes up to her again, but the female remains in a recumbent position, and the male moves away.

7 July 1981. 16.00 h. I find an adult female (Guy), a sub-adult female, and the immigrant male, Nicholas, in the river bed near Kousaunt windmill. The adult female is being followed by the others. The male comes up to the adult female and tries to mount her, at which the sub-adult female runs over and attacks him, knocking him off. The male

again attempts to mount, but the female continues walking. When the male mounts the female again, she stands, and the male stays on for almost a minute, resting his chin on her back. They then lie down, the two females lying together, the male lying some 50 m away. After half an hour the adult female starts chewing a bone. The sub-adult female stands up with an erection, licks the adult female on the back, then rests her chin on her back, and attempts to mount her (Fig. 7.6). The adult female soon moves off, followed by the other two. The male repeatedly approaches then backs away from the adult female, intermittently trying to mount with greater or lesser success, mainly depending on whether the sub-adult female intervenes or not.

Later they meet up with other members of the clan, including Hans, the central immigrant male. The females chase Nicholas briefly, but he continues to follow them at a distance until they lie down. After 16 min Hans moves over towards Nicholas, lowing (Table 5.6), and Nicholas backs away. Then Nicholas approaches the oestrus female, who is lying next to the sub-adult female, and sniffs her on the back. Hans ignores him, but the sub-adult female stands up, and the male backs away. He again approaches the female, but backs away when the sub-adult female stands up. This sort of behaviour continues for an hour until 23.30 h when Nicholas moves away. I leave at 02.30 h with no sign of Nicholas.

The male's fear of the female, the lack of cooperation on her behalf, and the lack of overt aggression between competing males are also characteris-

Figure 7.6. A sub-adult female spotted hyaena attempts to mount an adult female.

tic of mating observations on spotted hyaenas by Kruuk (1976), Henschel (1986), and Frank (1986b), and, except for the first characteristic, of those on brown hyaenas (section 7.2.1). The intervention by the sub-adult female was similar to Kruuk's observation of a ten-month-old cub chasing a male away from its mother, and the male-like sexual behaviour of the sub-adult female should be noted. The fact that the peripheral immigrant male (Nicholas) mated with a female, whereas the central immigrant male (Hans) who was present later, showed no interest in this female, is noteworthy, as Hans was probably her father (Appendix E). However, this mating did not produce cubs.

7.3.2   Immigrant males

Although immigrant males are mainly responsible for mating in spotted hyaena clans, they are not dominant, and undergo a prolonged period of assimilation before they are able to mate. Their behaviour is also unusual in that competition between males is usually not expressed in high levels of aggression.

7.3.2.1   ASSIMILATION

'An intruder will eventually become accepted if it persists in returning, hanging around the den and coming in on kills. This may, however, take a long time and complete acceptance may be achieved only after several months' (Kruuk 1972: 253).

Soon after I commenced observations on the Kousaunt clan in January 1979, two adult males (Jonas and Hans) were involved in an aggressive interaction:

31 January 1979. 18.00 h. Five spotted hyaenas, comprising two adult males, two sub-adult males, and an adult female, are at a wildebeest carcass at Groot Brak windmill. One of the adult males (Jonas) is standing submissively in the drinking reservoir. The others are standing around it, the second male (Hans) in particular, displaying a tendency to attack (Table 5.1). After a few minutes Hans jumps into the reservoir and attacks Jonas, biting at his neck, while the others look on with much interest (Fig. 7.7). The attacked animal roar-growls and some of the others grunt-laugh (Table 5.6). After a few seconds Hans leaves him. Jonas has blood on his neck, and remains in the reservoir with his tail curled under his belly and his ears flattened backwards. Ten minutes later Hans attacks again, with the others again showing interest. The attack is repeated three more times in the next hour. In between attacks Jonas remains in the reservoir, while the others feed on the carcass, or stand looking at him.

By 19.30 h six more members of the clan have arrived, and they take no more notice of the one in the reservoir. By 00.45 h he and Hans have disappeared.

Figure 7.7 Two immigrant male spotted hyaenas fight while other members of the clan look on.

After this Jonas became nomadic (Fig. 4.13). Hans remained in the area, but initially was infrequently seen in association with other clan members (Fig. 7.8). After a year there was a significant increase in the proportion of occasions in which he foraged with other members of the clan (5). Furthermore, after July 1979 other clan members ceased to chase him (cf. Nicholas below), and by June 1980 he would even occasionally sniff a female on meeting, although he only twice carried out the full meeting ceremony with one of them. Hans had probably joined the Kousaunt clan in January 1979, and gradually became assimilated into the clan. He may have displaced Jonas, or they may have arrived simultaneously, and competed for a place in the clan.

The next immigrant male (Nicholas) arrived in the Kousaunt territory in May 1981. His behaviour, and that of the clan towards him, repeated the events when Hans arrived. However, Nicholas was never accepted to the same degree as Hans, and he remained socially peripheral. From May 1981 until August 1983 Nicholas was only seen foraging with other members of the Kousaunt clan in five (5.1%) of 98 foraging groups. Nevertheless, he would frequently follow a foraging group at a distance. An example:

3 July 1982, 18.30 h. A group of nine hyaenas from the Kousaunt clan, comprising four adult females, the immigrant adult male, Hans, two sub-adults, and two large cubs leave the den. As they set off Nicholas appears some 60 m away, and the entire group chase him a short

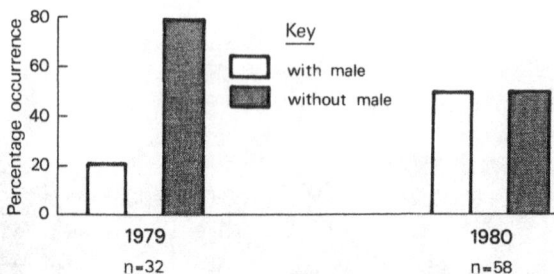

Figure 7.8   The percentage occurrence of foraging groups from the spotted hyaena Kousaunt clan with and without the immigrant male, Hans, in 1979 and 1980.

distance. The group continues on its way, and after 3 km I briefly see Nicholas following them at a distance of 100 m. After the group has moved another 30 km, mainly in a direction away from the den, Nicholas is again seen and briefly chased by them. Soon after this they rest for the day.

After they have moved 8 km the next evening, they chase a gemsbok herd. During the unsuccessful hunt an adult female becomes separated from the group. Nicholas appears, and follows her for 4 km until she lies down, when he disappears. Half an hour later the rest of the group appear, Nicholas again being seen in the background, and they move on. After 15 km they come to a windmill and once again Nicholas is seen. He is chased away, particularly by Hans, closely supported by the adult females. As they move off, Nicholas comes up to a sub-adult male who is lagging behind and they greet. The sub-adult then moves off to catch up with the others, and Nicholas continues to follow them at a distance, having already followed them for 60 km.

7.3.2.2   FORAGING

Once they have become assimilated into a clan, immigrant males play an important role in foraging, frequently leading a foraging group (Fig. 7.9). However, the immigrant male, Hans, foraged relatively more often on his own than expected, compared to the other individuals (6), and Nicholas too, usually foraged on his own.

When spotted hyaenas make a kill, there follows a period of rapid eating by all the animals present, before a more orderly sequence of feeding based on dominance and kinship is manifest (section 3.7.1.2). Of 13 kills at which an immigrant male was present from the start, he was the first to leave the carcass, apparently voluntarily, on ten (77%) occasions, a sub-adult being the first to leave in the remaining three. Once an immigrant male left a carcass, or if he arrived after the kill was made, or when a scavenged carcass was found, he had little chance of feeding while adult females were present (Fig. 7.10). If he attempted to do so he was chased away by a female, or even a sub-adult. However, immigrant males were sometimes allowed to feed with sub-adults. When the adult females left a carcass the immigrant males usually did too, even though there was food left over and the male had not eaten. It seems to be important for the immigrant males to stay in the females' company.

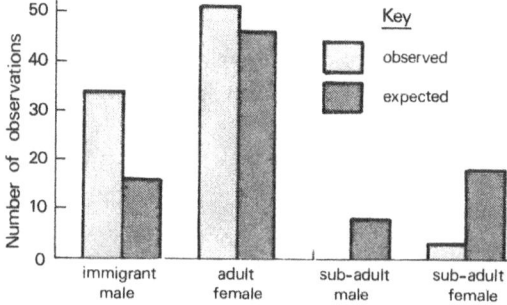

Figure 7.9 Observed and expected frequencies with which spotted hyaenas of different social categories led foraging groups. Expected values were calculated from the number of individuals in each social category in the clan.

Figure 7.10 An immigrant male spotted hyaena is prevented from feeding on a scavenged hartebeest by other clan members.

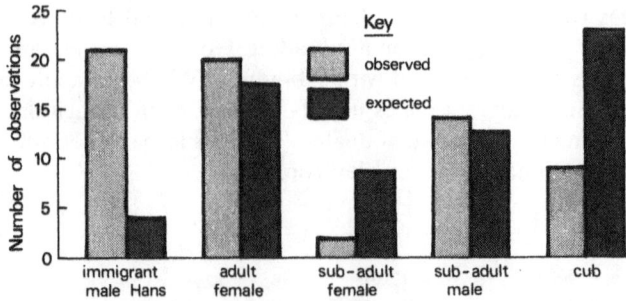

Figure 7.11 Observed and expected frequencies with which spotted hyaenas of different social categories from the Kousaunt clan chased the peripheral immigrant male, Nicholas. Expected frequencies calculated as in Figure 7.9.

7.3.2.3 RELATIONS BETWEEN IMMIGRANT MALES

Considering the stakes involved, i.e. access to several breeding females, relationships between competing immigrant male spotted hyaenas are restrained. In the only physical clash between immigrant males observed (section 7.3.2.1), no serious wounds were inflicted. However, Hans, the central immigrant male, was more likely to chase away the intruding immigrant male, Nicholas, than were any of the other group members (Fig. 7.11):

11 April 1982. Dunes west of Kousaunt. At 02.45 h five hyaenas, the immigrant male, Hans, three adult females, and a sub-adult male, kill an adult gemsbok. After feeding for an hour Hans becomes restless and starts to move about, lowing (Table 5.6). He pastes three times, then looks off to the east and repeatedly whoops while pawing the ground. The others continue to feed. After 5 min the other hyaenas look off to the east and all move off in that direction, tails up and lowing, towards Nicholas some 40 m away. They run after him a short distance, but then all except Hans return to the carcass. Soon after this another adult female from the clan arrives and proceeds to feed.

In the next 35 min Nicholas is chased away seven times by Hans, and twice by the sub-adult male, each time he approaches the carcass. Then one of the adult females moves off with a piece of the carcass, followed by the other two original adult females and Hans. The sub-adult male and the last adult female to arrive remain. Nicholas approaches to within 5 m of the carcass, picking up some scraps, and although he is ignored by the two feeding hyaenas, he does not feed on the carcass. At 06.00 h Nicholas moves away.

7.3.2.4 MALE REPRODUCTIVE STRATEGIES AND SUCCESS

As with brown hyaena males, there appear to be three reproductive options open to male spotted hyaenas, although the advantages and disadvantages of each differ between the two species.

1. Remain with natal clan. Mating opportunities are absent for natal males for the same reasons as for brown hyaenas. Neither can the male increase his inclusive fitness through kin selection by alloparental care, as guarding of young and provisioning them with food are absent in the spotted hyaena (section 6.4). Under certain conditions, there may be some advantage for a male spotted hyaena to delay his departure from his natal clan, for example, in a high density situation where a possible vacancy in a neighbouring clan can be monitored from the safety of the natal clan (Henschel & Skinner 1987), but this did not appear to be a profitable option in the southern Kalahari.

2. Become nomadic. Unlike with brown hyaenas, this does not seem to be a good option, probably because spotted hyaena males need to go through an extended period of assimilation before they are accepted by females.

3. Immigrate. If a male can become accepted into a clan and maintain his position he will gain numerous mating opportunities. Obviously this is the best option.

Spotted hyaenas have a polygynous mating system (Frank 1986b, Henschel & Skinner 1987, this study). Polygynous societies are characterized by sexual dimorphism where males tend to be larger and more aggressive than females, compete with other males, and display to females. This is associated with the development of weaponry and adornments (Partridge & Halliday 1984, Keverne 1985). Spotted hyaenas clearly do not fit this model; males are generally smaller than females and are dominated by them, they possess no specialized weaponry and adornments, they appear to exhibit little overt aggression to each other, and they do not display to females in a conspicuous manner. Frank (1986b) believed that spotted hyaena females do not exercise choice, leaving the males to sort out mating rights among themselves. I contend that the females may exercise a measure of mate choice by determining which males are allowed to join their clan. There are a number of studies in several species demonstrating female choice (Majerus 1986).

Persistence seems to be an important quality for a male spotted hyaena to possess. Males invest time and energy in following adult females, both prior to and on being integrated into a clan (see also Frank 1986b, Henschel & Skinner 1987), and in reinforcing their presence by whooping and pasting (Ch. 5). This is best illustrated by the behaviour of the immigrant male, Nicholas, who spent 27 months on the periphery of the Kousaunt clan, repeatedly being rejected by other clan members. His persistence was apparently rewarded when the clan split, and he departed with the Two-Two coalition (section 7.3.3.3), his behaviour immediately changing from that of a peripheral to a central immigrant male. This behaviour of immigrant males may be analogous to the conspicuous displays of males to attract females found in other species.

Therefore, rather than inter-male competition expressed in a high degree of overt aggression being the key to male reproductive success in the spotted hyaena, more subtle mechanisms of inter- and intrasexual selection such as the persistence of the males seem to be prevalent. The reason for this unusual system may be the fact that females dominate males.

Recently, R. Goss (personal communication) observed two male spotted hyaenas in northern Botswana team up to attack a third in a fierce fight, which left the loser severely injured, and apparently resulted in one of the victors becoming the central immigrant male of the clan. So under certain conditions interactions between males can escalate dramatically. The fact that two males joined forces to attack a third is particularly intriguing.

7.3.3   Adult females

A spotted hyaena clan revolves around its adult females. They are both numerically and socially dominant to the males, and take the lead in most clan activities. Amongst the females and their offspring too, there are intricate relationships.

7.3.3.1   DOMINANCE AND REPRODUCTIVE SUCCESS

The concept of dominance has been discussed in section 7.2.3. It is of importance amongst spotted hyaena females as dominance is frequently associated with mating rights (Schaller 1972, Wilson 1975, Halliday 1978, Kaufman 1983, Frank 1986b). The results of dyadic disputes over food between the three females of the Kousaunt clan who were adults for the entire study (Fig. 4.6), suggest that Olivia was the dominant one, followed by Ella, and then Two-Two (Table 7.7).

As cooperation between relatives in raising young was virtually absent in spotted hyaenas during the study, the number of surviving descendants which accrued to an individual provides a measure of its success in passing on its genetic material during this time to future generations (Clutton-Brock *et al.* 1982). During the period January 1978 to December 1983,

Table 7.7   Outcomes of dyadic competitive interactions over food between three of the spotted hyaena females from the Kousaunt clan.

Loser	Winner			
	Olivia	Ella	Two-Two	Total losses
Olivia	-	1	0	1
Ella	7	-	1	8
Two-Two	3	3	-	6
Total wins	10	4	1	-

shortly after the clan had split (section 7.3.3.3), each of the three females mentioned above produced five cubs which survived until they were at least two years old (Fig. 4.6). However, in terms of their reproductive success there were differences between Olivia and the other females (Fig. 7.12). Olivia produced three female cubs and Ella and Two-Two two each. Olivia's two oldest female cubs, Goldie and Guy, were recruited into the Kousaunt clan, and both had bred by December 1983 (Fig. 4.6). By October 1985, her third daughter had also produced cubs (M. Knight personal communication). Furthermore, Olivia's son, Silver, successfully immigrated to the St John's clan (sections 4.2.3, 4.2.4), where he had mating opportunities. In this connection it is of interest that Frank et al. (1985) recorded an unusually high androgen level, for a natal male, in the son of a dominant female in the Mara. Frank (1986b) postulated that the sons of dominant females, because of their inherited status and high level of aggression, are more likely to become successful breeders than are the sons of low ranking females.

Neither of Ella's two female cubs were recruited into the Kousaunt clan. They disappeared when they were three years and three years nine months old respectively, before they had bred. The chances of them being able to join another clan were poor (see section 4.2.4).

Four of Two-Two's offspring left the clan with her (Fig. 4.6), and, although it is possible that the females subsequently bred (section 7.3.3.3), the number of descendants accruing to Two-Two between January 1978 and December 1983 were less than to Olivia. Furthermore, Two-Two's oldest male cub had not joined another clan by September 1983 when he was nearly five years old.

As she grew up, Olivia's oldest daughter, Goldie, began to dominate the other females of the clan, even challenging and sometimes beating her mother. The younger daughter, Guy, on the other hand, only occasionally challenged her half-sister (Goldie) and her mother, but always lost (Fig. 7.13). Frank (1986b) maintained that a female's position in the hierarchy

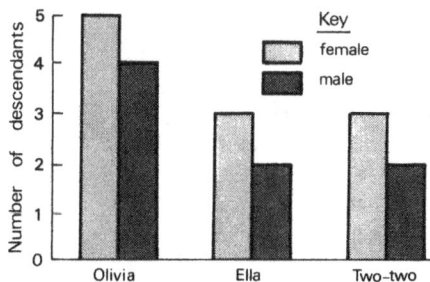

Figure 7.12 Reproductive success of three female spotted hyaenas from the Kousaunt clan, measured as the number of descendants of each female which survived to two years of age between 1978 and 1983.

was determined by its mother's rank. This held for Goldie, but not for Guy. Goldie produced cubs more rapidly than the other females, even when intervals between litters which died are excluded, whereas Guy had a relatively poor reproductive performance (Table 7.8). There was, in fact, a tendency for females of high social rank to produce cubs at shorter intervals than subordinate females (7). Similar results were obtained by Frank (1986b).

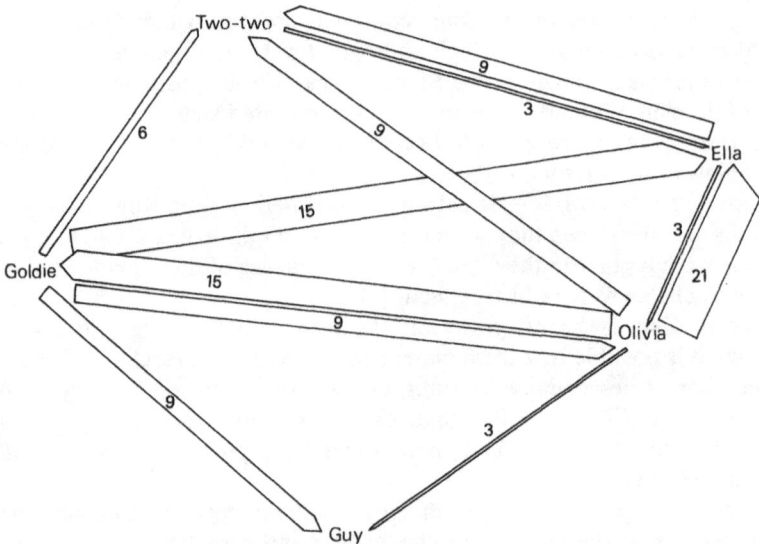

Figure 7.13 Outcomes of dyadic aggressive interactions between the adult female spotted hyaenas of the Kousaunt clan, 1982–1984. The arrows point from winner to loser.

Table 7.8   Intervals in months between litters of cubs raised to sub-adulthood by spotted hyaena females from the Kousaunt clan, arranged in descending order of dominance.

Female	Inter-litter interval	Mean
Olivia	18, 19, 17	18.0
Goldie	15, 16	15.5
Ella	25, 18, 18	20.3
Two-Two	20, 19	19.5
Guy	22	22.0

Overall mean 18.5 ± 0.8; $n = 11$

7.3.3.2 RELATEDNESS AND FORAGING GROUPS

Although the female, Guy, was low ranking, she and other subordinate animals, i.e. sub-adults and cubs, were tolerated at carcasses by their more dominant close relatives, whereas more distantly related animals were not. Some examples:

> 3 December 1982. 21.55 h. Olivia, and Two-Two's sub-adult daughter, Soap, arrive at an adult gemsbok carcass on which Olivia's adult daughter, Guy, is feeding. Olivia starts feeding with Guy, but Soap remains in the background. After a few minutes Soap comes to the carcass, but is attacked by Olivia and retreats. At 22.23 h Olivia's oldest daughter, Goldie, and her 14-month-old son, Archer, together with Ella's sub-adult son, Curley, arrive. Their arrival is accompanied by much vocalizing and all the animals, including Soap, feed briefly amidst some squabbling. Then Soap and Curley leave the carcass, but the others feed with little squabbling. At 22.52 h Soap's litter-mate, the sub-adult male, Joe, arrives. For the next hour he, Soap, and Curley lie or circle around a few metres from the carcass while Olivia and her descendants feed.

> 28 February 1983. 21.15 h. Olivia and her two six-month-old cubs, her adult daughters, Goldie and Guy, the adult female, Ella, and Two-Two's sub-adult daughter, Soap, are lying at a den in the river bed near Bedinkt. Suddenly, Goldie, Guy, and Soap jump up and run 400 m to a fresh springbok carcass on which some jackals are feeding. The jackals scatter and Goldie and Guy immediately start feeding, but Soap keeps away. Olivia arrives, and the three adult females feed on the carcass, with Soap in the background. After 5 min Olivia's two cubs arrive, and, although they do not join in the melee, they pick up some scraps next to the carcass, approaching closer to the feeding hyaenas than does Soap. Then Ella's and Goldie's sub-adult sons, Longchri and Archer respectively, arrive. Longchri makes no attempt to approach the carcass, but Archer, after chasing Soap a short distance, comes close and picks up some scraps. However, he is chased away when he tries to join the females. Half an hour after arriving at the carcass, Guy carries the skin and bones back to the den at which Ella is still lying, having made no attempt to join the others.

This tendency for close relatives to receive preference at carcasses is further illustrated by the fact that the degrees of relatedness between the members of 32 observed feeding groups from the Kousaunt clan were higher than the degrees of relatedness between the members of 32 equal-sized feeding groups randomly generated by a computer from the pool of available hyaenas in the clan (8). If close relatives do receive preference at carcasses, it would be expected that closely related individuals forage together. The degrees of relatedness between individuals in

131 observed foraging groups of Kousaunt hyaenas, and 28 groups of Kaspersdraai animals, were again higher than equal numbers of equal-sized random groups from each clan, chosen as above (9). This may explain why Kalahari spotted hyaena foraging groups are often larger than they need be for maximum foraging efficiency (section 3.6.10). Large groups are not always needed to overcome prey, but the amount of food available at the end of the hunt is often enough to feed several individuals.

7.3.3.3 COALITIONS AND CLAN FISSION

The number of females in the Kousaunt clan reached a peak early in 1983 (Fig. 4.7), as three coalitions developed in the clan. Each coalition comprised one of the three original adult females and their offspring. The coalition of the dominant female, Olivia, was the largest, comprising nine individuals of all ages, followed by the low-ranking Two-Two coalition of five individuals, and the middle ranking Ella coalition of four individuals (Fig. 4.6). The proportion of mixed coalition foraging groups to single coalition foraging groups encountered declined in 1983 compared with 1982 (10). Furthermore, from May 1983 onwards, members of the Two-Two coalition were rarely seen at the den, the matriarch, Two-Two, and her eldest daughter, Soap, never being seen there. Nor were they seen interacting amicably with members of the other two coalitions.

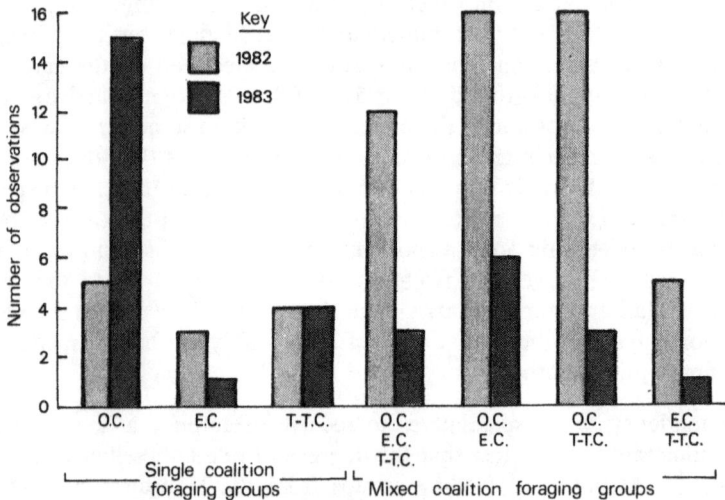

Figure 7.14 Frequencies with which at least one member of one of the three spotted hyaena coalitions in the Kousaunt clan foraged with at least one other hyaena from its own coalition (single coalition foraging group), or at least one other hyaena from another coalition (mixed coalition foraging groups) in 1982 and 1983. O.C. = Olivia coalition, E.C. = Ella coalition, T.T.C. = Two-Two coalition.

In late July 1983 Olivia and Guy sustained leg injuries. So severe were Guy's injuries that she was still dragging her hind legs noticeably six months later. A clue as to how these injuries were sustained, and a significant observation with regard to the fission of the clan, is recorded below:

31 July 1983. The entire Olivia coalition (Fig. 4.6), and Ella's six-month-old cub, leave the den at 19.17 h for Kousaunt windmill 2 km away. When they arrive at the windmill they lie down. Ten minutes later the five members of the Two-Two coalition, plus the immigrant male, Nicholas, come running in from the southeast. The members of the Olivia coalition scatter, the cubs retreating, the larger animals running over to the approaching hyaenas. A bloody battle involving all the animals except the cubs and the immigrant male ensues. In amongst the dust and confusion it is difficult to make detailed observations, but the hyaenas are biting each other around the face and underparts, with the attacked individuals going down on their stomachs to protect themselves (Fig. 7.15). The interaction is accompanied by snarls, roar-growls, yells, hoot-laughs, squeals, and giggles (Table 5.6), with the cubs standing some 30 m away, whooping. Once, one of the hyaenas attempts to take shelter under my vehicle, but others pull it out. After 9 min the members of the Two-Two coalition break away and run off, being chased by the other animals, whooping. After approximately 1 km they give up the

Figure 7.15 An intra-clan fight between the members of two coalitions of the Kousaunt spotted hyaena clan.

chase and after 3.4 km the chased animals stop. All of them, but particularly Two-Two and her two eldest offspring, the male, Joe, and the female, Soap, are covered in blood around their faces and fore-quarters. The immigrant male, Nicholas, is unscathed. After 70 min the hyaenas move off slowly in a northeasterly direction. They visit six latrines, at which they social sniff, but only deposit three pastings.

The next evening at sunset they are back at the den. Ella and her cub are also there, but the two groups ignore each other. Soon after sunset the Two-Two animals and Nicholas move off in an easterly direction.

This was the last occasion that any members of the Two-Two coalition, and Nicholas, were seen in the Kousaunt territory. In December 1984 they were found at a den 40 km to the southeast of the Kousaunt clan's den (M. Knight personal communication) (Fig. 4.9). The Olivia and Ella coalitions continued to forage separately after the clan had split (Fig. 7.14), although the members of the two coalitions denned together. In 1986 Ella left the Kousaunt clan (M. Knight personal communication).

A similar clan fission may have occurred in the Kaspersdraai clan when three adult females left the area together, and established themselves in the St John's area (sections 4.2.3, 4.2.4).

Clan fission in the spotted hyaena is similar to troop fission in certain primates (Nash 1976, Chepko-Sade & Sade 1979), in that fission occurs along genetically related lines and the subordinate groups break away.

7.3.3.4 FEMALE REPRODUCTIVE STRATEGIES AND SUCCESS

The three options open to female spotted hyaenas are:
1. Remain with natal clan. The number of animals in a clan, and, therefore, the number of adult females, is likely ultimately to be limited by the food supply (section 4.5). Accordingly, not all females can stay with their natal clan as adults. The female offspring of dominant females are more likely to be recruited than the offspring of lower ranking females. Unlike males, female spotted hyaenas which stay with their natal clan may be able to enhance their fitness through kin selection, by forming breeding coalitions with their daughters, and allowing close relatives access to carcasses. This latter behaviour may be important for the survival of large cubs and sub-adults which are not yet proficient at hunting.
2. Immigrate. Low-ranking females may occasionally be able to join another clan individually, although the conditions under which this occurs are unknown.
3. Establish a new clan. Females of low rank may leave the clan with other closely related individuals, and establish a new clan in a vacated area. The success of this strategy will depend on the availability of a suitable area, a situation which, as with brown hyaenas, is unlikely to be found often in a saturated environment.

The amount of parental investment required by spotted hyaena females to raise their young successfully appears to be higher than it is in any other carnivore. The energetic costs of lactation are high (Pond 1977), and for the first year of their lives spotted hyaena cubs rely mainly on milk. This is an unusually long suckling period for a carnivore (section 6.2.3). It may explain the relatively large size of females in comparison with males (Bertram 1979) – the 'big mother' hypothesis of Ralls (1976).

Female dominance and reproductive success were related in the present study, as was also indicated by Frank (1986b), and as has been found in a number of primate species (Drickamer 1974, Dunbar & Dunbar 1977, Silk *et al.* 1981). This was not due to the fact that dominant females raised more young, but because their daughters gained breeding opportunities within the clan. The daughters of lower-ranking females either had to emigrate alone to an uncertain future, or delay breeding until the clan split. The fact that the dominant female of the Kousaunt clan produced the only male (Silver) who was known to join another clan, may also be of relevance. In fact, Frank (1986b) postulated that dominant females will produce more sons than daughters, although I found no evidence to support this statement. Frank also suggested that the sons of the dominant female were able to delay their departure from their natal clan until they reached adult size and sexual maturity, whereas the sons of subordinate females were forced to leave sooner in search of less competitive feeding. In contrast, in my study, the son (Silver) of the dominant female was, at 30 months, the youngest male to leave a clan, and the son of the low-ranking Two-Two was, at 44 months, the oldest (Fig. 4.6). Perhaps this is an example of different strategies being optimal under different conditions.

7.3.4   Inter-clan relations

Although territories overlap, interactions between different spotted hyaena clans are rare in the southern Kalahari. Only one confrontation between the members of two clans was observed (section 5.1.2). On two other occasions hyaenas reacted to the whoops of those from a different clan by turning around and moving away from the callers and in one of these cases scent was also believed to have played a role (section 5.2.2.1).

But what may be the most significant observations on inter-group relations, those of infanticide, are clouded by a lack of completeness. Once, the spoor of four members of the Seven Pans clan came to the Kaspersdraai clan's den, in which there were four two- to three-month-old cubs. Two days later the cubs had moved to a new den 6 km away, and one of them was missing. Could the Seven Pans hyaenas have killed the cub? On another occasion a nine-month-old cub from the Kousaunt clan sustained an injury to the front foot and appeared to have been bitten in the shoulder. At the same time this cub's litter-mate and a one-year-old cub disappeared. One month later two six-month-old cubs from the same clan sustained similar injuries, although both survived. Kruuk (1972) recorded strange spotted hyaena adults killing cubs in the Serengeti, and Henschel (1986) has circumstantial evidence for spotted hyaenas from neighbouring clans killing cubs in Kruger National Park.

Although the disappearance and injuries of the cubs in the present study could have been caused by predators such as lions, it is possible that they were inflicted by neighbouring spotted hyaenas. It does not seem that immigrants were responsible for the Kousaunt cubs' disappearances and injuries, as sightings of nomadic males in a territory did not appear to be correlated with cub mortality, and the intensity with which the immigrant male, Nicholas, was chased away from dens was no different from when he was chased away at any other time. It is possible, therefore, that infanticide in spotted hyaenas is a manifestation of inter-clan aggression, similar to what has been recorded in coyotes (Camenzind 1978). This is in contrast to lions and some primate species where infanticide is carried out by immigrant males after taking over a pride or troop (Bertram 1975, 1976, Hrdy 1979).

7.4   Evolutionary trends in the social systems of the two species

'Unravelling the evolution of any social system must begin with an understanding of the roots of female behavior, since the behavior of males is largely adapted to that of females' (Wrangham & Rubenstein 1986: 469). The most simple social unit observed in the Hyaenidae is that of an adult female and her latest litter of cubs inhabiting a territory (Table 4.1). The female presumably mates with a nomadic male. This is similar to the primitive carnivore social system postulated by Kleiman and Eisenberg (1973). However, where the dispersion pattern of food allows it (and in habitats where spotted hyaenas can survive this may always be the case), brown hyaena and spotted hyaena females are able to form larger groups, When this happens it may become worthwhile for males to join the groups, as there will be several females with which to mate. Although superficially the differences in social organization between the two species appear great, with brown hyaenas regarded as solitary and living at low densities, and spotted hyaenas as social and often occurring at high densities, in reality the social units are the same; a varying number of related females with which associate a varying number of males. The main differences in social organization between the two species are that spotted hyaenas often forage in groups, and brown hyaenas practise alloparental care. The reasons for these differences can be found in the different feeding habits of the two species.

Wrangham (1980) suggested that competition for clumped, high quality food resources that are defendable has led to the evolution of female-bonded groups in primates. Hyaenas (and most other large social carnivores) fit this model. Spotted hyaenas, particularly, often experience intense competition at food, perhaps more so than other carnivores (Frank 1986b), and certainly more so than brown hyaenas. Spotted hyaenas, like certain primates (Wrangham 1980), and unlike other social carnivores, form coalitions within groups so that partners gain an even larger share of the resources that a group already controls, leading to enhanced fitness for these animals.

7.5 Summary

	Brown hyaena	Spotted hyaena
Average degree of relatedness (r) between clan members	0.26	0.31
Mating system	Polygynous, mating mainly by nomadic males	Polygynous, mating mainly by immigrant males
Mating behaviour	Numerous copulation attempts of short duration	Numerous copulation attempts of short duration Male shows fear of female
Dominance hierarchy	Absent	Females dominate immigrant males and form hierarchical coalitions
Male reproductive options	Become nomadic Immigrate Remain with natal clan and possibly increase fitness through alloparental care	Immigrate Become nomadic (rare)
Female reproductive options	Stay with natal clan and breed Set up own territory	Stay with natal clan and breed if mother is dominant Split off with descendants to form new clan Immigrate as solitary (rare)
Infanticide	Not recorded	Probably by neighbours
Social organization	Female-bonded groups	Female-bonded groups with coalitions within the group

Statistical tests

1. Comparison of the frequencies with which strange male brown hyaenas engaged in sexual behaviour as opposed to non-sexual behaviour with group-living females versus the frequencies with which group-living males did so.
$\chi^2 = 19.89$; d.f. $= 1$; $p < 0.001$; $N = 75$.

2. Comparison of the frequencies with which strange males and group-living males were observed in the Kwang brown hyaena territory in months that mating was observed versus the frequencies they were observed in months when no mating occurred.
$\chi^2 = 6.34$; d.f. $= 1$; $p < 0.02$; $N = 27$.

3. Comparison of the frequency with which male brown hyaenas carried food to the den versus the frequency with which females did so.
$\chi^2 = 1.00$; d.f. $= 1$; $p > 0.05$; $N = 41$.

4. Comparison of the frequency with which brown hyaenas of the same sex from neighbouring clans acted agonistically towards each other versus the frequency with which animals of opposite sex from neighbouring clans did so.
Fisher exact probability test: $p = 0.005$; one-tailed; $N = 10$.

5. Comparison of the frequencies with which the spotted hyaena immigrant male, Hans, foraged alone and with other clan members in 1979 versus 1980.
$\chi^2 = 7.69$; d.f. $= 1$; $p < 0.01$; $N = 90$.

6. Comparison of the observed frequency with which the spotted hyaena immigrant male, Hans, foraged alone versus the expected frequency if he foraged alone in the same frequency as the other adults in the clan.
$\chi^2 = 4.29$; d.f. $= 1$; $p < 0.05$; $N = 19$.

7. Correlation between the social rank of female spotted hyaenas of the Kousaunt clan and the interval between successfully raised litters.
Spearman rank correlation coefficient: $r_s = 0.8$; $p > 0.05$; $N = 5$.

8. Comparison of the degrees of relatedness between the members of 32 observed feeding groups of spotted hyaenas from the Kousaunt clan that were permitted to feed at a given carcass versus the degrees of relatedness between the members of 32 equal-sized feeding groups randomly generated by a computer from the pool of available hyaenas in the clan.
Kolmogorov–Smirnov one-sample test: $D = 0.088$; $p < 0.01$; two-tailed.

9. Comparison of the degrees of relatedness between the members of (i) 131 observed foraging groups of spotted hyaenas from the Kousaunt clan, and (ii) 28 observed foraging groups from the Kaspersdraai clan versus the degrees of relatedness between the members of 131 and 28 equal-sized feeding groups randomly generated by a computer from the pool of available hyaenas in the two clans respectively.
(i)   Kolmogorov–Smirnov one-sample test: $D = 0.227$; $p < 0.01$; two-tailed.
(ii) Kolmogorov–Smirnov one-sample test: $D = 0.149$; $p < 0.01$; two-tailed.

10. Comparison of the frequencies with which spotted hyaenas from the Kousaunt clan were observed to forage in single and mixed coalition groups in 1982 versus the frequencies they did so in 1983.
$\chi^2 = 14.21$; d.f. $= 1$; $p < 0.01$; $N = 94$.

8 Relations between, and management considerations for, brown hyaenas and spotted hyaenas

The ecological and behavioural relationships between the two hyaena species are complex (Mills & Mills 1977), and have probably had a major influence on their distribution and abundance. Where they are sympatric these relationships must be considered when formulating management strategies for the two species. In presenting some guidelines for their management, I do so with the firm conviction that all indigenous animals are resources of inherent interest and value to humanity, and that we have a moral obligation to conserve them. This must be balanced with many practical considerations which, when we are dealing with carnivores, are particularly complex.

8.1 Relations between brown hyaenas and spotted hyaenas

8.1.1 At food

In Chapter 2 it was shown that the diets of brown hyaenas and spotted hyaenas in the southern Kalahari are very different. The brown hyaena is predominantly a scavenger of vertebrate remains, supplementing its diet with wild fruits, insects, and birds' eggs. The spotted hyaena is predominantly a hunter–scavenger of large and medium-sized mammals. However, there is a measure of overlap in their diets, particularly regarding scavenged carcasses. When large and medium-sized carcasses are concerned, brown hyaenas stand to lose a significant amount of food to spotted hyaenas. On the other hand, brown hyaenas may gain from scavenging from spotted hyaenas' kills.

Competition for carcasses between the two species varied between different areas (Table 8.1). In the Kwang area brown hyaenas usually found a carcass first, and were usually able to feed on it uninterrupted by spotted hyaenas, losing only 18% of their carcasses to spotted hyaenas. On the other hand, in the Kousaunt area spotted hyaenas were usually first onto a carcass, and in those few cases that brown hyaenas were known to locate a carcass first, they were displaced by spotted hyaenas.

Table 8.1   Relative frequencies with which brown hyaenas and spotted hyaenas were first to locate
a carcass in two areas, and the number of occasions that brown hyaenas were displaced
by spotted hyaenas. The Kwang area is between Nossob camp and Bedinkt, the
Kousaunt area between Bedinkt and Leijersdraai (see Fig. 1.1).

	Kwang area	Kousaunt area	χ^2
Carcass found by brown hyaena first	22	5	15.17; d.f.=1; $p<0.001$
Carcass found by spotted hyaena first	4	16	
Brown hyaena displaced by spotted hyaena	4	5	8.87*; d.f.=1; $p<0.01$

* With Yates' correction

In the five cases that spotted hyaenas displaced brown hyaenas in the
Kousaunt area (Table 8.1), a single brown hyaena and one to five spotted
hyaenas were involved. In three of the observations the brown hyaena ran
away at the approach of the spotted hyaenas, but in two it stood its ground
as three and five spotted hyaenas approached it. In both these cases one of
the spotted hyaenas grabbed the brown hyaena by the side of the neck and
dragged and shook it for 1–2 min, with the brown hyaena snarling and
yelling and trying to bite back. The other spotted hyaenas ran around the
struggling pair, but did not join in the attack. When the brown hyaena
managed to free itself, the spotted hyaenas left it and went to the food. On
both occasions blood was drawn on the neck of the brown hyaena.

In the four cases that brown hyaenas were displaced by spotted hyaenas
in the Kwang area, the interactions were more complex, and once, a
spotted hyaena failed to displace a brown hyaena. Four times, one spotted
hyaena, and once, two, were involved. In all cases there were from two to
four brown hyaenas involved at varying stages of the interactions. Two
examples are given below:

17 January 1980. 21.33 h. Kwang windmill. A brown hyaena is feeding
on a wildebeest carcass and another is lying down 30 m away. A spotted
hyaena walks towards the carcass, but stops 10 m away. The brown
hyaena raises the hair on its back and hoot-laughs (Table 5.6), scraping
the ground with its forefeet, and the spotted hyaena moves away.
Twelve minutes later the spotted hyaena comes back to the carcass, and
both species feed together for a few seconds. Then the brown hyaena
breaks a small piece off and moves away with it. It returns five minutes
later with the spotted hyaena still feeding, circles the carcass 5 m away,
and then departs.

Five minutes later the second brown hyaena comes to the carcass, and
the two species feed together (Fig. 8.1) for 3 min until the spotted
hyaena moves away. It returns 2 min later, but the brown hyaena

Figure 8.1 A brown hyaena and a spotted hyaena feed together briefly on a scavenged wildebeest.

continues to feed, hair raised and snarling. The spotted hyaena lies down 30 m away. After 35 min the brown hyaena moves off. Sixteen minutes later the spotted hyaena returns to the carcass, feeds for 19 min, then moves away as well. Two minutes later a lion comes to the carcass.

19 May 1978. 23.55 h. Dunes near Kwang. A spotted hyaena comes running up to a hartebeest carcass on which two brown hyaenas are feeding. The brown hyaenas move away slowly, and the spotted hyaena starts to eat, the brown hyaenas circling around some 20–50 m away. Five minutes later a second spotted hyaena comes to the carcass.

At 00.25 h one of the brown hyaenas moves close to a feeding spotted hyaena. As it approaches with its hair raised, the spotted hyaena curls its tail over its back and raises the hair on the back of its neck (Fig. 8.2). It then moves towards the brown hyaena, followed closely by the second spotted hyaena. The brown hyaena retreats a few metres then stops, crouches slightly, and with its hair still raised, gives a short deep growl (Table 5.6). The spotted hyaenas stop 1–2 m from the brown hyaena, and remain in that position for 20 s, looking off in different directions, while the brown hyaena growls intermittently. They then move back to the carcass and continue feeding, and the brown hyaena moves away. A few minutes later the brown hyaena again approaches the spotted hyaenas and the same thing happens. After this the brown hyaena lies down 15 m from the carcass.

Figure 8.2 A brown hyaena approaches a spotted hyaena feeding on a hartebeest it had previously taken over from two brown hyaenas.

During the night, two more brown hyaenas come to the vicinity of the carcass. One or other of them intermittently approaches the spotted hyaenas with the same results as described above. At 05.00 h one of the brown hyaenas comes up to the carcass, snarling, with its hair raised. The two spotted hyaenas move away at a short distance, and the brown hyaena starts feeding. The spotted hyaenas come back to the carcass, and one of them drags it away from the brown hyaena. Three minutes later the brown hyaena again approaches. As it does so one of the spotted hyaenas drags the carcass off for 10 m, and continues feeding, but the second one moves away. The brown hyaena again approaches the feeding spotted hyaena, which then moves away after the other one, leaving the remains of the carcass to the brown hyaenas.

In these interactions the spotted hyaenas were dominant to the brown hyaenas most of the time. Yet I gained the impression that by persistently disturbing the spotted hyaenas, the brown hyaenas caused them to abandon the food sooner than they might otherwise have done. This behaviour was less dramatic and persistent than that of spotted hyaenas towards feeding lions (section 2.6.2.1), or black-backed jackals towards feeding brown hyaenas (setion 2.6.1.4), but its function appears to be the same, i.e. to cause the larger carnivore to abandon the carcass. However, spotted hyaenas never lost a significant amount of food.

The amount of food available to brown hyaenas from the remains of

Table 8.2   Amount left over from large and medium sized spotted hyaena kills in two areas. Areas as defined in Table 8.1.

	Kwang area	Kousaunt area	χ^2 *
A few scattered bones	3	29	
Skeleton and skin	4	8	10.85; d.f. = 1; $p < 0.001$
Skeleton with some meat	3	0	

* A few scattered bones vs. skeleton and skin + skeleton with some meat

spotted hyaenas' kills also varied between the two main study areas (Table 8.2). When the spotted hyaenas made a kill in the Kousaunt area, other clan members quickly joined them on the kill which they consumed completely. However, when some members of the clan moved into less well utilized areas of the clan's territory, only those hyaenas which participated in the hunt fed from the carcass, and the adult females in particular were likely to return to the den early, thus often leaving substantial remains.

8.1.2   Away from food

4 January 1980. 07.25 h. Kwang windmill. Seven spotted hyaenas come to the windmill and chase a brown hyaena which is drinking there. After 500 m the brown hyaena backs up against a tree, sitting up on its carpals, with its hair raised, ears back (as in Fig. 8.3), mouth open, and snarling. Within a few seconds the spotted hyaenas move away slightly, and stand looking around in different directions. The brown hyaena moves off slowly for a few metres, then starts to run. Immediately it does so the spotted hyaenas run after it again. After 100 m the brown hyaena stops in the open, adopting the same posture as earlier. The spotted hyaenas run around it. Twice, one of them darts in and nips it in the lower back, and once, one sniffs at the brown hyaena's back briefly. After a few seconds they retreat for 10–20 m, and stand looking off in different directions (Fig. 8.3). The brown hyaena slowly backs up to a nearby bush facing the spotted hyaenas all the time, then, after 1–2 min, it moves off slowly with its hair raised. Some of the spotted hyaenas watch it go, but they do not chase it again.

On 28 occasions a single brown hyaena interacted with a varying number of spotted hyaenas away from food – in 19 (68%) cases one or two spotted hyaenas were involved, in 7 (25%) cases three to five were involved, and in one case each (7%), seven and nine spotted hyaenas were involved. These interactions are summarized in Figure 8.4 and clearly show the dominance

Figure 8.3 One of seven spotted hyaenas which have cornered a brown hyaena stands and looks away from the brown hyaena.

of spotted hyaenas over brown hyaenas, and that most interactions between these two species result in harassment, even death, for the brown hyaena.

In spite of the usual harassing experience for brown hyaenas when they meet up with spotted hyaenas, there is at times a mutual attraction between them. In 6 (21%) of their meetings, the brown hyaena approached closer to the spotted hyaenas, as if inviting an attack (although in all six cases there were only one or two spotted hyaenas present), and the two species stood or circled each other at close quarters. Although the spotted hyaenas could easily have attacked the brown hyaena, they did not. A similar relationship has been described between spotted hyaenas and striped hyaenas (Kruuk 1976).

Although brown hyaenas sometimes stand and face spotted hyaenas, they appear to be frightened by the whoop call. Once, two brown hyaenas were feeding on a carcass when a spotted hyaena came running up to them. One of the brown hyaenas stood its ground, but the other ran off. The two species fed on the carcass for a short time, then the spotted hyaena moved away. Twenty minutes later the brown hyaena was still feeding on the carcass when a spotted hyaena whooped close by. Immediately, the brown hyaena ran away. On six occasions I tested a resting brown hyaena's reaction to a tape-recording of a spotted hyaena whooping. Each time the brown hyaena showed concern; twice it ran away, twice it walked over 1 km and twice 50 m away before lying down again.

28
Brown hyaena detects spotteds

4
Moves away before spotteds
detect it

24
Moves closer

6
They circle each other,
then part

18
Chased by spotteds

5
Outruns spotteds

13
Spotteds catch up

4
Spotteds surround it, but
do not attack

9
Spotteds attack

8
Brown escapes

1
Brown killed

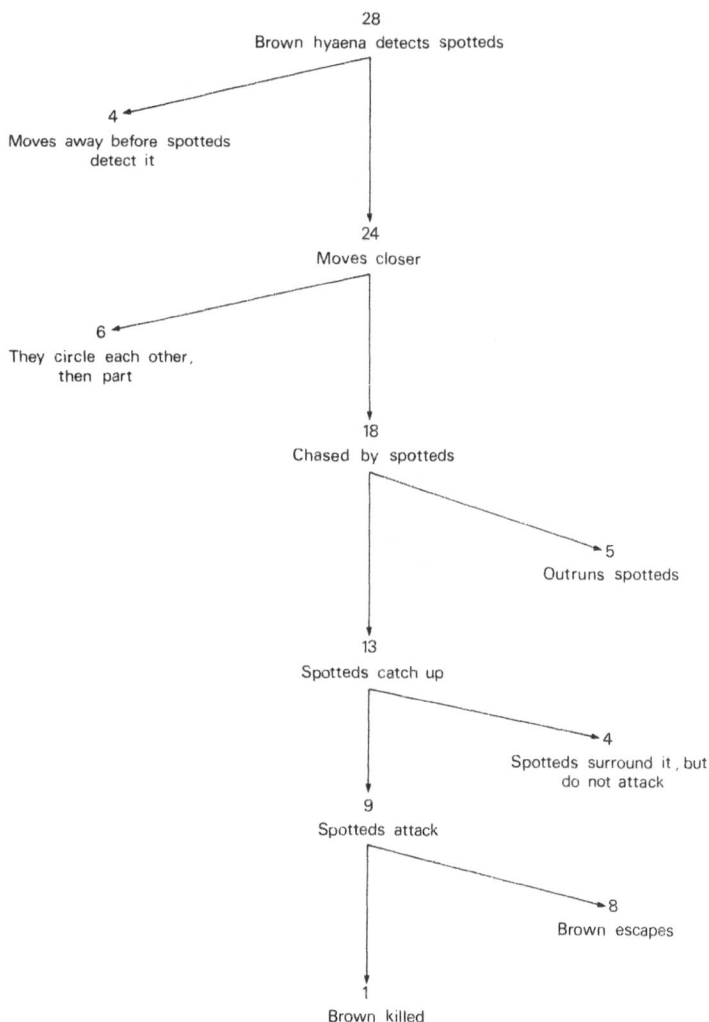

Figure 8.4   Interactions between brown hyaenas and spotted hyaenas away from food.

Spotted hyaenas also raid brown hyaena dens. On one occasion, two spotted hyaenas came running up to a den, having obviously sensed it from 2–3 km away, at which three brown hyaena cubs of six months old were lying. The cubs darted into the den and the spotted hyaenas went in after them. However, they were not able to reach the cubs because of the narrow tunnel. Then a fourth cub of approximately one year old was seen approaching the den. The spotted hyaenas chased it and after 150 m caught up with it. They disappeared behind a bush, with the brown hyaena cub growling and yelling. After a few seconds the cub reappeared, moving on its carpals, with the spotted hyaenas over it and mouthing it. The brown hyaena rolled free and darted into a minor den (see section 6.1) unharmed. The spotted hyaenas returned to the den, sniffed around, chewed on a few bones, and four more times went down into the mouth of the den. Twelve minutes after arriving at the den they moved off.

On three occasions spotted hyaenas scavenged at a brown hyaena den. Once, seven ate the remains of a carcass, and twice they ate a few bones. Usually brown hyaena cubs take food into the den, reducing losses of food.

8.1.3   Impact of spotted hyaenas on brown hyaenas

Spotted hyaenas are obviously dominant to brown hyaenas. They deprive them of some food, often harass them away from food, and even, occasionally, kill a brown hyaena. On the other hand, brown hyaenas obtain some food from spotted hyaenas by scavenging from their kills. What is the balance of this relationship in the southern Kalahari?

I compared the relative numbers of the two species seen when driving along the Nossob river bed at night in the Kousaunt and Kwang study areas (Table 8.3). The data show that brown hyaenas were more common in the Kwang area than in the Kousaunt area, and vice versa for spotted hyaenas. The habitat in these two regions was similar, so that habitat differences do not seem to be the reason for differences in density. For the reasons outlined in sections 8.1.1 and 8.1.2 it is doubtful that brown hyaenas could have adversely affected spotted hyaena numbers in the Kwang area – the higher density of spotted hyaenas in the Kousaunt area was a result of spotted hyaenas usually denning in this area (Fig. 6.2). However, spotted hyaenas may have affected brown hyaena numbers in the Kousaunt area. Not only was there probably less food available to brown hyaenas there, because of competition for scavenged carcasses with spotted hyaenas, but brown hyaenas may have avoided the area to avoid contact with spotted hyaenas.

Several other studies on closely related, sympatric carnivores have also shown that the smaller species tend to be more common in areas not

Table 8.3   Comparison of the frequencies with which brown hyaenas and spotted hyaenas were encountered at night in two areas along the Nossob river bed. Areas defined as in Table 8.1.

	Kwang area $n = 1053$km	Kousaunt area $n = 1166$km	χ^2
Brown hyaenas	20	6	26.40; d.f. = 1; p < 0.001
Spotted hyaenas	10	50	

frequented by the larger ones. In the Namib Desert brown hyaenas are mainly found along the coast, whereas spotted hyaenas range further inland (Skinner & Van Aarde 1981). Kruuk (1976) recorded that in East Africa striped hyaenas were more common in low-productivity habitat areas, which tended to be avoided by spotted hyaenas. Schaller (1967), and Seidensticker (1976), found that leopards tended to avoid high density tiger areas in India and Nepal, and Berg & Chessness (1978), and Fuller & Keith (1981), found that coyotes tended to avoid areas frequented by wolves in North America.

Brown hyaenas are able to survive on widely scattered, small food items, whereas spotted hyaenas are dependent on large and medium-sized ungulates. Because of the nomadic nature of much of their prey, and the consequent long periods of scarce food availability, spotted hyaenas can only survive in the southern Kalahari at a low density (0.8–$1.0/100\,\mathrm{km}^2$). The overall influence of spotted hyaenas on the brown hyaena population in the southern Kalahari, therefore, is small. At higher spotted hyaena densities it is reasonable to suppose that the negative influence on the brown hyaena population will increase. This may explain why, for example, in the Kruger National Park, with a higher spotted hyaena density than the southern Kalahari (7–$20/100\,\mathrm{km}^2$), brown hyaenas are extremely rare (Mills 1985b). It is of interest to note that in the Kruger National Park spotted hyaenas exploit a wider niche, scavenge more, and hunt less, than in the southern Kalahari (Henschel 1986, personal observations).

8.2   Management considerations

8.2.1   Management in the southern Kalahari

The management objective in the Kalahari Gemsbok National Park is to preserve the Park as a functional part of the southern Kalahari ecosystem. Any management considerations for hyaenas must be compatible with this overall management strategy.

The brown hyaena population in the southern Kalahari is viable and healthy because of the large area of suitable habitat. The low density of other large carnivores, particularly spotted hyaenas, probably favours the brown hyaena.

Certain aspects of the brown hyaena's behaviour seem to be poorly adapted to competition from other large carnivores. Brown hyaena cubs spend long periods foraging on their own, making them vulnerable to attack. Adults carry food, sometimes strong-smelling carcasses, long distances back to their cubs, inviting interception by other carnivores. The presence of these carcasses at the den may also attract predators. It is logical that the higher the densities of other large carnivores, the greater will be the chances that foraging cubs, food-carrying adults, and dens will be attacked.

Judging by the speed and extent to which the size of the Kwang brown hyaena clan and its territory fluctuated (Ch. 4), the southern Kalahari brown hyaena population has the potential to fluctuate appreciably and rapidly, depending on the food supply, particularly the availability of ungulates. These fluctuations should be seen as part of the dynamic nature of an ecosystem.

The spotted hyaena population in the southern Kalahari is low. The nomadic nature of the large ungulates is the limiting factor. Strangely enough, spotted hyaenas do not appear to have reacted to the establishment of sedentary wildebeest populations along the two dry river beds due to the provision of water by means of boreholes (Eloff 1966, Mills & Retief 1984b). The area along the Nossob river bed between Nossob camp and Bedinkt (Fig. 1.1), which had the largest concentration of wildebeest along the entire Nossob river bed (unpublished observations), had relatively few spotted hyaenas (Table 8.1).

Although spotted hyaenas are able to inhabit areas with little surface water, they may be more dependent on water than are brown hyaenas (Skinner & Van Aarde 1981, Tilson & Henschel 1986). Spotted hyaenas appear to be rarer in the central Kalahari than in the southern Kalahari (Smithers 1971, Owens & Owens 1978), where there is no permanent water, even though it is a more productive region. The provision of water, therefore, may have increased spotted hyaena numbers in the southern Kalahari.

Spotted hyaenas are efficient hunters of gemsbok calves. It is unfortunate that more reliable figures for the number of gemsbok are not available, so that the impact of spotted hyaena predation on the gemsbok population could have been determined more accurately. Nevertheless, the indications are that in spite of low hyaena numbers this is significant (section 2.5). An increase in the number of spotted hyaenas without an increase in the number of gemsbok, such as could be brought about by the further provision of water, would likely put more pressure on the gemsbok population (and the brown hyaena population).

There is some evidence that rabies may depress the southern Kalahari spotted hyaena population (section 4.1.5). It is not known whether rabies is indigenous to the Kalahari (R. Bengis and V. de Vos personal communication). However, the facts that semi-feral dogs are responsible for the majority of epizootics in most of Africa, and that the southern Kalahari falls in the area of overlap between the jackal/semi-feral dog and viverrid rabies areas (Meredith 1982), suggests that the disease's prevalence is at least partly due to human-induced factors. Furthermore, unlike some diseases, rabies does not select weakened individuals. For these reasons there may be a case for countering this disease by immunization (Macdonald 1980b). However, at this stage it would seem to be more important to gain more information about rabies in the southern Kalahari. Specifically, all field staff in the area should be taught and prepared to collect samples from suspected rabid animals. If rabies is found, the spotted hyaena population should be carefully monitored in order to document the effect of the epizootic on the population.

Boreholes and rabies notwithstanding, the most important consideration for the continued well-being of brown hyaenas and spotted hyaenas, as well as the entire southern Kalahari ecosystem, is that the two national parks continue to be managed as a single ecological unit with the minimum of intervention. Without the large-scale movements of ungulates between the Kalahari Gemsbok and Gemsbok National Parks it will be impossible to maintain the ungulate populations. Not only would this seriously affect the food supply for hyaenas, it would serve as the death knell for this priceless gem, one of Africa's last large pristine ecosystems.

8.2.2   Management in other parts of southern Africa

8.2.2.1   BROWN HYAENA

Besides the southern Kalahari, viable brown hyaena populations exist in several other conservation areas: the Central Kalahari Game Reserve, Botswana (Owens & Owens 1978), the coastal regions of the southern Namib Desert (Skinner & Van Aarde 1981), and probably Kaokoland (Viljoen 1980), the Etosha National Park in northern Namibia (Von Richter 1972), and Iona National Park in southern Angola (Huntley 1974). An area in excess of $1000\,km^2$, where competition with spotted hyaenas is slight, would appear to be required for the successful maintenance of a viable brown hyaena population. No other suitable conservation areas within the brown hyaena's distribution range exist at present.

The introduction of brown hyaenas to smaller game reserves is a complex issue. Although they would not affect the ungulate populations, as was found to be the case with an active predator such as the cheetah (Pettifer 1981a, b), it might become necessary to artificially feed them in order to keep them to the confines of the reserve. This could be turned to an advantage if arrangements could be made for tourists to observe the animals at bait stations. Such a confined population would also pose overpopulation problems and would be similar to conserving the species in a zoological garden (Stuart et al. 1985).

Outside conservation areas, good habitat for brown hyaenas exists on agricultural land – there is an adequate supply of food from domestic animals which die of natural causes, human refuse, and wild animals, and spotted hyaenas and other large carnivores are usually absent. Viable populations of brown hyaenas exist in some of these areas, particularly in parts of the Transvaal Province of South Africa (Skinner 1976), and Botswana (Smithers 1971).

Certain areas of South Africa have been defined as suitable only for extensive cattle production (Anon. 1965). These could be designated as brown hyaena conservation areas. Here, a major research effort should be aimed at finding effective means for the rational management of brown hyaenas (Stuart *et al*. 1985). The magnitude of loss of domestic livestock to brown hyaena predation must be established, and, where necessary, attempts made to find economically efficient control methods, with the emphasis on non-lethal, or selective lethal methods (Sterner & Shumaker 1978, Wade 1978). Predator control seeks to decrease predator populations, but the aim should be to lessen the damage, not necessarily the pests themselves (Andelt 1987).

There are ways to reduce predator damage besides increasing predator mortality (Giles 1978, Andelt 1987). There are several management practices which might protect livestock from predation, and which should be tested (see Andelt 1987 and references therein). Synchronizing calving, the use of portable electric fencing for enclosing breeding herds at night, increased vigilance by shepherds during the lambing season, the use of livestock guard dogs, frightening devices such as strobe lights and sirens, and taste aversion conditioning (Gustavson *et al*. 1974, Gustavson & Nicolaus 1987), have all been used to greater or lesser effect in coyote control in North America. In addition, research is needed on how farmers can obtain the optimum ecological benefit from brown hyaenas; for example, how best to deal with the carcasses of domestic animals that die from disease.

The support of the people living in the relevant areas is essential if such a plan is to be successfully implemented, a point emphasized by Mech (1977), and Macdonald & Boitani (1979), when discussing wolf management. This would require an efficient and extensive education campaign aimed at the local residents.

8.2.2.2 SPOTTED HYAENA

The spotted hyaena enjoys a much wider distribution than the brown hyaena (section 1.4.1). In southern Africa, beside the southern Kalahari, viable spotted hyaena populations occur in the Kruger National Park (Mills 1985b) and neighbouring private game reserves (in the Transvaal) (Bearder 1977, personal observations); the Hluhluwe–Umfolosi Complex (Whateley 1981, Whateley & Brooks 1978) and Mkuzi Game Reserve (Deane 1962) in South Africa; Etosha National Park (Joubert & Mostert 1975) and the Namib Desert (Skinner & Van Aarde 1981) in Namibia; over much of northern Botswana (Smithers 1983); in Hwange and Ghona-re-Zhoa National Parks and the Zambezi valley in Zimbabwe (Smithers & Wilson 1979); and in Mozambique (Smithers & Lobao Tello 1976). It is also conserved in many national parks and game reserves in the rest of sub-Saharan Africa.

Outside conservation areas the status of the spotted hyaena is precarious. This is particularly so in South Africa where sophisticated agricultural practices have practically exterminated large predators. It is difficult to reconcile the presence of spotted hyaenas in these areas, and there is little point in attempting to do so. In less developed agricultural areas the emphasis should again be on damage control. Kruuk (1980) suggested that the loss of livestock to spotted hyaena predation in Kenya could be prevented by better vigilance during the day, by preventing animals from straying, and by shutting livestock in bomas at night.

8.2.3 Conclusions

Because of its nocturnal and secretive behaviour it is difficult to establish the exact status of the brown hyaena in southern Africa. Undoubtedly its numbers have been reduced in many parts of its range in recent years, particularly in those areas where small domestic livestock are farmed. However, there are several large conservation areas in ideal brown hyaena habitat which harbour viable brown hyaena populations. Furthermore, the species adapts easily to many human activities. As long as the large conservation areas are maintained in their present state, and if a rational approach to the management of brown hyaenas in other areas can be taken, the future survival of the brown hyaena can be viewed with optimism.

The spotted hyaena is unable to inhabit agricultural areas as successfully as is the brown hyaena. Neither is it as well adapted to the arid regions of southern Africa. However, in several of the larger conservation areas in the higher rainfall regions of Africa the spotted hyaena is numerically the dominant large carnivore. Even more than the brown hyaena, the spotted hyaena's future is tied to the long-term future of these conservation areas.

If this book has helped in any way to convince people that hyaenas are worth conserving, not only for their intrinsic value, but because of their beauty and fascinating behaviour, it will have been worth the many hours of toil that it has taken me to try to convert the wonderful experiences of watching them, into some sort of coherent and scientifically meaningful form.

8.3 Summary

1. Spotted hyaenas dominate brown hyaenas at food and in certain areas may deprive brown hyaenas of food. Less frequently, brown hyaenas are able to scavenge from spotted hyaena kills.
2. Away from food, spotted hyaenas frequently harass brown hyaenas, occasionally killing them. In spite of this there is often a mutual attraction between the two species.
3. Spotted hyaenas may have a detrimental effect on brown hyaena numbers in certain areas. However, because of the low numbers of spotted hyaena in the southern Kalahari, this effect is small in this area.

4. The brown hyaena population in the southern Kalahari is healthy and viable. The spotted hyaena population is less secure. The habitat is marginal because of the nomadic nature of the ungulates. In addition, rabies may be holding the population slightly below the carrying capacity. More needs to be learnt of the effect of rabies on the population.

5. The most important consideration for the continued well being of the entire southern Kalahari ecosystem is to continue to manage the two national parks as one ecological unit.

6. It is possible to manage and conserve brown hyaenas in certain agricultural areas, but spotted hyaenas can only be effectively protected in large conservation areas.

Appendix A
Common and scientific names
of species mentioned in
the text

Plants

Fungae
  Kalahari truffle   *Terfezia pfelii*

Monocotyledonae
  Buffalo grass   *Panicum coloratum*
  Gha grass   *Centropodia glauca*
  Dune grass   *Stipagrostis amabilis*
  Small bushman grass   *Stipagrostis obtusa*
  Giant stick grass   *Aristida meridionalis*
  Speckled vlei grass   *Eragrostis bicolor*
  Lehman's love grass   *E. lehmanniana*
  Feather-top chloris   *Chloris virgata*
  Kalahari sour grass   *Schmidtia kalahariensis*

Dicotyledonae
  Camelthorn   *Acacia erioloba*
  Grey camelthorn   *Acacia haematoxylon*
  Candle acacia   *Acacia hebeclada*
  Blackthorn   *Acacia mellifera* subsp. *detinens*
  Desert date   *Balanites aegyptiaca*
  Shepherd's tree   *Boscia albitrunca*
  Rhigozum   *Rhigozum trichotomum*
  Monechma   *Monechma incanum*
  Tsama (melon)   *Citrullus lanatus*
  Gemsbok cucumber   *Acanthosicyos naudianus*
  Brandy bush   *Grewia flava*
  Merremia   *Merremia tridentata*

Mammals

Order Pholidota
  Pangolin   *Manis temminckii*

Order Lagomorpha
  Hare   *Lepus* spp.

Order Rodentia
  Porcupine   *Hystrix africaeaustralis*
  Springhare   *Pedetes capensis*
  Ground squirrel   *Xerus inauris*

Order Carnivora
Wolf	*Canis lupus*
Coyote	*Canis latrans*
Black-backed jackal	*Canis mesomelas*
Golden jackal	*Canis aureus*
Red fox	*Vulpes vulpes*
Cape fox	*Vulpes chama*
Bat-eared fox	*Otocyon megalotis*
Arctic fox	*Alopax lagopus*
African wild dog	*Lycaon pictus*
Wolverine	*Gulo gulo*
European badger	*Meles meles*
Honey badger	*Mellivora capensis*
Striped polecat	*Ictonyx striatus*
Tiger	*Panthera tigris*
Lion	*Panthera leo*
Leopard	*Panthera pardus*
Cheetah	*Acinonyx jubatus*
Caracal	*Felis caracal*
African wild cat	*Felis lybica*
Striped hyaena	*Hyaena hyaena*
Brown hyaena	*Hyaena brunnea*
Spotted hyaena	*Crocuta crocuta*
Aardwolf	*Proteles cristatus*
African civet	*Civettictis civetta*
Small-spotted genet	*Genetta genetta*
Yellow mongoose	*Cynictis penicillata*

Order Pinnipedia
| Cape fur seal | *Arctocephalus pusillus* |

Order Tubulidentata
| Aardvark | *Orycteropus afer* |

Order Perissodactyla
| Black rhinoceros | *Diceros bicornis* |
| Burchell's zebra | *Equus burchelli* |

Order Artiodactyla
Hippopotamus	*Hippopotamus amphibius*
Giraffe	*Giraffa camelopardalis*
Red deer	*Gervus elaphus*
Moose	*Alces alces*
Blue wildebeest	*Connochaetes taurinus*
Red hartebeest	*Alcelaphus buselaphus*
Common duiker	*Sylvicapra grimmia*
Springbok	*Antidorcas marsupialis*
Thomson's gazelle	*Gazella thomsonii*
Steenbok	*Raphicerus campestris*
Impala	*Aepyceros melampus*
Grey rhebok	*Pelea capreolus*
Sable	*Hippotragus niger*
Gemsbok	*Oryx gazella*
Buffalo	*Syncerus caffer*
Kudu	*Tragelaphus strepsiceros*
Bushbuck	*Tragelaphus scriptus*

Eland	*Taurotragus oryx*
Defessa waterbuck	*Kobus defassa*

Birds
Ostrich	*Struthio camelus*
Secretary bird	*Sagittarius serpentarius*
White-backed vulture	*Gyps africanus*
Kori bustard	*Ardeotis kori*
Red-crested korhaan	*Eupodotis ruficrista*
Black korhaan	*Eupodotis afra*
Crowned plover	*Vanellus coronatus*
Cape dikkop	*Burhinus capensis*
Ant-eating chat	*Myrmecocichla formicivora*

Insects
Harvester termite	*Hodotermes mossambicus*
Snouted harvester termite	*Trinervitermes* sp.

Appendix B
Estimated numbers of some ungulates in the spotted hyaena study area.

Estimated numbers of some ungulates in the spotted hyaena study area, the Kalahari Gemsbok National Park plus a 30km strip along the Nossob river bed in the Gemsbok National Park, 1979-1983. Two counts per year were carried out, one at the end of the rains, one in the middle of the dry season. River bed counts were carried out from the ground, dune counts from the air. These figures are not accurate and are only given as an indication of the actual numbers and relative amounts of fluctuation that occurred. Totals are not necessarily the sum of river bed and dunes, as maximum and minimum counts were not always made in both habitats at the same time.

Species	Habitat	Mean	Max	Min	SD	CV
Gemsbok	River bed	270	748	58	229	85%
	Dunes	12 011	18 093	5247	4240	35%
	Total	12 281	18 155	5528	4215	34%
Wildebeest	River bed	734	1265	212	423	58%
	Dunes	18 175	170 840	207	53 645	295%
	Total	18 909	172 027	816	53 805	285%
	Total-Aug '79 count	1896	2685	816	722	38%
Eland	River bed	0	0	0	0	0%
	Dunes	3263	13 099	12	4697	144%
	Total	3263	13 099	12	4697	144%
Hartebeest	River bed	260	859	4	303	117%
	Dunes	3607	17 967	253	5301	147%
	Total	3867	18 014	259	5233	135%
Springbok	River bed	4441	7359	929	2230	50%
	Dunes	2135	5625	413	1700	80%
	Total	6576	11 326	1342	3038	46%

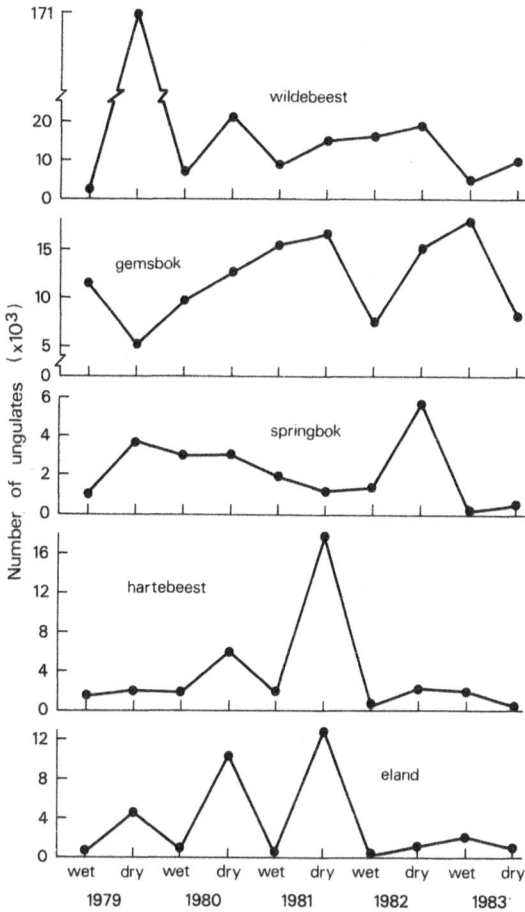

Figure B.1 Estimated numbers of some ungulates in the spotted hyaena study area during each of the twice-annual counts, 1979–1983.

Appendix C
Ageing criteria of ungulates based on eruption of teeth in bottom jaws and tooth wear

Age-class	Gemsbok	Wildebeest	Hartebeest	Eland	Springbok
I	Up to M1 just erupting.	Up to M1 just erupting.	Up to M1 just erupting.	Up to M1 just erupting.	Up to M1 just erupting.
II	M2 just erupting.	M2 just erupting.	M2 just erupting.	M2 just erupting.	M1 fully erupted.
III	M3 just erupting.	M3 just erupting.	M3 just erupting.	M3 just erupting.	M2 fully erupted.
IV	Up to full permanent dentition with no tooth wear.	Up to full permanent dentition with no tooth wear.	Up to full eruption of permanent dentition with no tooth wear.	Up to full permanent dentition with no tooth wear.	M3 erupting.
V	Little tooth wear. Incisor length (of chewing surface) >2× width. Inf. of PM3 and PM4 still open. Ectostylid of M3 not worn.	Incisor length > 2× width. Inf. of PM3 and PM4 usually open.	Incisor length > 2× width. Inf. of PM3 and PM4 usually open.	From fully erupted dentition with all inf. present to post. inf. of M3 gone.	Full permanent dentition. Ant. inf. of M1 gone.
VI	Tooth wear visible. Incisor length < 2× width. Inf. of PM3 usually closed. Ectostylid of M3 worn to oval (but not joined to chewing surface).	Incisor length < 2× width, oval and often touching. Inf. of PM3 usually closed. Post inf. of PM4 if present closed at post. end.	Incisor length > 2× width. Ant. inf. of PM3 still open; post inf. closed or gone. Ant. on PM4 closed or gone.	Ant. inf. of M1 gone.	Post. inf. of M1 gone. Ant. inf. of M2 worn.

VII	Inf. on PM4 gone, at least on one side. Ant. inf. of M1 U-shaped. Ectostylid of M3 worn to circle.	Incisors oval or rectangular, often not touching. Post inf. of PM4 gone. Ant. inf. of M1 small and oval, or round; post. inf. U-shaped.	Incisors round, sometimes not touching. Ant. inf. of PM3 usually open. Ant. inf. of M1 small and oval, or round, or gone; post. inf. U-shaped.	Post. inf. of PM4 gone. Both inf. of M1 gone.	Ant. inf. of M2 gone; post. inf. worn.
VIII	Incisors oval or round, still touching. Ant. inf. of M1, small and oval, or round, or gone. Post. inf. U-shaped.	Some incisors worn to round and not touching. Ant. inf. of M1 gone; post. inf. small and oval, or round, or gone.	Incisors round, not touching. Ant. inf. of PM3 and PM4 present or absent. Ant. inf. of M1 gone; post. inf. small and oval, or round, or gone. Ant. inf. of M2 U-shaped to round.	Ant. inf. of M2 and M3 gone.	Post. inf. of M2 gone. Inf. of M3 worn.
IX	Incisors often not touching, some non-functional. Post. inf. of M1 small and oval, or round, or gone.	Incisors not touching, some non-functional. Post. inf. of M1 gone. Inf. of M2 worn.	Ant. inf. on M2 small and oval, or round, or gone; post. inf. U-shaped, or slightly worn.	Post. inf. of M2 gone.	All inf. except sometimes middle one on M3 gone.

It was not feasible to establish the absolute ages of these ungulate age-classes as methodological problems of achieving this; i.e. sectioning, staining and counting growth rings from the teeth of known aged animals, as well as those in the sample, were beyond the scope of the study. From the work of Talbot and Talbot (1963) and Attwell (1980) on wildebeest, and Kerr & Roth (1970) and Jeffery & Hanks (1981) on eland, and assuming that eruption times for gemsbok and hartebeest are similar to these other two species, it was possible to assign approximate absolute ages to animals of these species with erupting teeth. Animals in age-class 1 were 0–6 months old, those in age-class 2 were 7–12 months old, those in age-class 3 were 13–26 months old, and those in age-class 4 were 27–40 months old. From age-class 5 onwards the teeth are fully erupted. For springbok, animals in age-class 1 were 0–3 months old, those in age-class 2 were 4–6 months old, those in age-class 3 were 6–16 months old, and those in age-class 4 approximately 18 months old (Rautenbach 1971, R. Liversidge pers. com.).

Appendix D
Methods used to measure territory sizes

Hyaena territories in the southern Kalahari are large, ill-defined areas. I arbitrarily decided that the minimum amount of data required to measure territory size for brown hyaenas was that animals from a particular clan must have been followed, either directly or from spoor, for at least 300 km over a period of at least four months. For spotted hyaenas the criteria set were that animals from a particular clan must have been followed for at least 1000 km over a period of at least one year.

Macdonald et al. (1980) advocated that where observations are adequate, home range boundaries should be drawn in by hand and measured with a planimeter. Using the criteria for each species referred to above as 'adequate' data, enough data were collected to measure six clan territories for each species (Tables D.1, D.2). Three of the brown hyaena territories are the Kwang clan's territory,

Table D.1   Territory sizes (km^2) for six brown hyaena clans using three measuring techniques.

Clan	Time period	Distance followed or tracked (km)	Convex polygon	Restricted polygon	Grid-square
Kwang 1	Dec 74 - March 75	500	317	297	258
Kwang 2	Jan 76 - July 76	1235	249	228	213
Kwang 3	July 77 - Aug 78	845	235	215	195
Kaspersdraai	Oct 72 - April 73	442	276	267	245
Rooikop	May 72 - Oct 74	302	424	381	284
Seven Pans	Jan 75 - Sept 75	539	481	461	319
Mean ± SE		644 ± 139	330 ± 41	308 ± 39	252 ± 19

Table D.2   Territory sizes (km^2) for six spotted hyaena clans using three measuring techniques.

Clan	Time period	Distance followed or tracked (km)	Convex polygon	Restricted polygon	Grid-square
Kousaunt 1	June 79 - Dec 80	1314	1293	989	794
Kousaunt 2	May 81 - Nov 82	1124	1103	1046	750
Kousaunt 3	Dec 82 - Feb 84	1131	705	553	538
Kaspersdraai	Sept 82 - Jan 84	1045	881	814	769
St. John's	Oct 82 - Feb 84	1049	1727	1394	1006
Seven Pans	Nov 82 - Feb 84	1098	2420	1776	1281
Mean ± SE		1132 ± 47	1355 ± 257	1095 ± 177	856 ± 104

measured over three time periods, with a gap of a year or more between each
(Table D.1). Three of the spotted hyaena territories are the Kousaunt clan's
territory, also measured over three time periods. There was a four-month gap in
fieldwork between the first (Kousaunt 1) and second (Kousaunt 2) time periods
(Table D.2), and a shift in emphasis of the area utilized (see Fig. 4.15), as well as
the beginnings of the fission of the clan (section 7.3.3.3), between the second and
third (Kousaunt 3) time periods.

   Three methods were used to measure each of the 12 territories (Tables D.1,
D.2). The first method employed the drawing of a convex polygon (Dalke 1942,
Mohr 1947) around the extremities of the observed movements (Figs D.1, D.2).
The second involved the restricted polygon method of Wolton (1985) (Figs D.1,
D.2). Here, a 6.25 km² grid system was placed over each territory and a fix was
defined as the centre of any grid square through which a hyaena moved (Fig. D.2).
The length of a line of the polygon may not exceed the average distance between all
the fixes and the arithmetic mean centre (i.e. the mean x and y coordinates of all
fixes) of the range. The third method employed grid cells on a map (Fig. D.2). The
number of cells, in this case the 6.25 km² grid squares, in which hyaenas were
observed to move were totalled and multiplied by the size of a grid square (Voigt &
Tinline 1980).

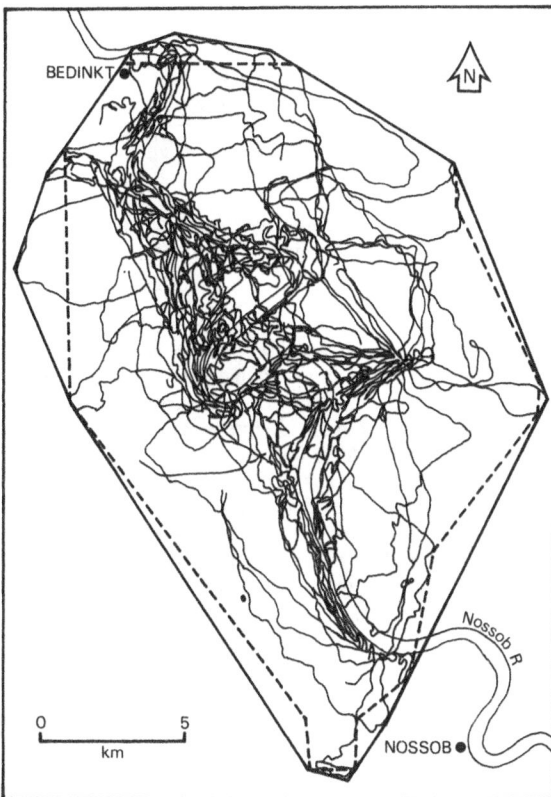

Figure D.1 The observed movements of brown hyaenas from the Kwang 2 clan, showing the
convex polygon (solid line) and restricted polygon (dotted line) boundaries.

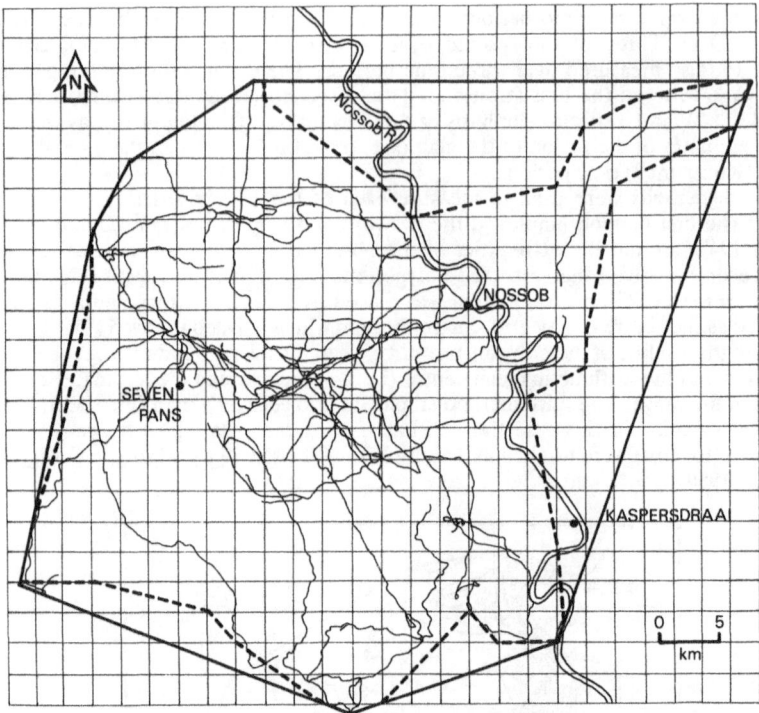

Figure D.2 The observed movements of spotted hyaenas from the Seven Pans clan, showing the grid system used and the convex polygon (solid line) and restricted polygon (dotted line) boundaries.

Measurements by the restricted polygon method were the ones used to test hypotheses regarding territory size. This method had the advantage over the minimum convex polygon method in that it does not include large areas of unused ground. What areas to exclude are decided in an objective manner, although this may not always be biologically meaningful. Although this was not a problem with the brown hyaena territories (see example in Fig. D.1), it was with the spotted hyaena territories, the extreme example of which is illustrated in Figure D.2. The grid square method on the other hand excludes areas that were most likely utilized by clan members, but because of the extremely large size of the territories were not detected by the sampling method. This is also illustrated in Figure D.2.

Appendix E
Degrees of relatedness
between clan members

1. Brown hyaenas

In calculating the degrees of relatedness (r) between the various members of the Kwang brown hyaena clan (Table E.1), the following assumptions were made:

1. Females preferentially suckled their own cubs.
2. Each litter of cubs was fathered by a different nomadic male, who was unrelated to the females in the clan and to the other nomadic males.
3. The male Hop-a-long was a natal male, related to the matriarch, Normali, by the mean of all the other relationships in the clan, i.e. 0.28.

Table E.1   Probable coefficient of relatedness between the members of the Kwang brown hyaena clan. Underlined individuals are females.

	Normali	Charlie	Chinki	Shimi	Sanie	Thunberg	Brunnea	Hyaena	Elvis	Bing	Lazarus	Mistly	Chrimoi	Hop-a-long
Normali	-	50	50	50	50	50	50	50	25	25	25	50	50	28
Charlie	-	-	50	25	25	25	25	25	25	25	25	25	25	14
Chinki	-	-	-	25	25	25	25	25	50	50	50	25	25	14
Shimi	-	-	-	-	50	25	25	25	13	13	13	25	25	14
Sanie	-	-	-	-	-	25	25	25	13	13	13	25	25	14
Thunberg	-	-	-	-	-	-	50	50	13	13	13	25	25	14
Brunnea	-	-	-	-	-	-	-	50	13	13	13	25	25	14
Hyaena	-	-	-	-	-	-	-	-	13	13	13	25	25	14
Elvis	-	-	-	-	-	-	-	-	-	50	50	13	13	07
Bing	-	-	-	-	-	-	-	-	-	-	50	13	13	07
Lazarus	-	-	-	-	-	-	-	-	-	-	-	13	13	07
Mistly	-	-	-	-	-	-	-	-	-	-	-	-	50	14
Chrimoi	-	-	-	-	-	-	-	-	-	-	-	-	-	14
Hop-a-long	-	-	-	-	-	-	-	-	-	-	-	-	-	-

2. Spotted hyaenas

In calculating the degrees of relatedness between the various members of the Kousaunt (Table E.2) and Kaspersdraai (Table E.3) spotted hyaena clans, the following assumptions were made:

1. Females suckled only their own cubs.
2. Where there was only one adult immigrant male in a clan at the time the cubs were conceived, he was their father. For the Kaspersdraai clan this held for the entire study, but in the Kousaunt clan there were two resident immigrant adult males in the clan from May 1981 until August 1983 (Fig. 4.6). However, from the behaviour of all but one of the females towards the second male (Nicholas), they would not have allowed him to mate with them. Accordingly, the male, Hans, fathered all the cubs born during the period he was resident

in the clan, with the exception of Guy's cub (Fig. 4.6), which could have been fathered by either male. This cub was excluded in analyses involving degrees of relatedness.

3. The father of the cubs born to Olivia and Two-Two in 1978 (Fig. 4.6), was the same individual.

4. The mating males were unrelated to each other and to the females in the clans in which they mated, except where father–daughter matings probably occurred.

5. The degrees of relatedness between the three key adult females of the Kousaunt clan, Olivia, Ella and Two-Two (Fig. 4.6), were the average of the known degrees of relatedness between the ten female pairs from the Kaspersdraasi clan, i.e. 0.38.

Table E.2   Probable coefficient of relatedness between the members of two spotted hyaena clans

Underlined individuals are females.

	Olivia	Ella	Two-Two	Goldie	Guy	Sophia	Stedul	Joe	Soap	Archer	Curly	Longchri	Beaky	Nie	Lorna	Rian	Wilf	Rosy	Lady Di	Paka	Hans	Nicholas
Olivia	-	30	30	50	50	15	15	15	15	25	25	15	50	50	25	25	15	15	15	15	0	0
Ella		-	30	20	20	50	50	20	20	10	50	50	20	20	20	10	20	20	50	50	0	0
Two-Two			-	20	20	20	50	50	10	20	20	20	20	20	10	10	50	50	20	50	0	0
Goldie				-	25	10	10	10	10	50	10	10	25	25	30	30	35	35	35	10	50	0
Guy					-	35	35	35	35	38	35	35	50	50	38	38	35	35	35	10	50	0
Sophia						-	50	33	33	30	50	50	33	33	30	30	33	33	50	10	50	0
Stedul							-	33	45	30	50	50	45	45	30	30	50	50	45	25	50	0
Joe								-	50	30	45	45	45	45	30	30	50	50	45	25	50	0
Soap									-	30	45	45	45	45	30	30	50	50	45	25	50	0
Archer										-	30	30	38	38	50	50	30	30	30	05	50	0
Curly											-	50	33	33	29	29	45	45	50	10	50	0
Longchri												-	45	45	30	30	45	45	45	10	50	0
Beaky													-	50	38	38	45	45	45	10	50	0
Nie														-	50	50	45	45	45	10	50	0
Lorna															-	50	30	30	30	05	50	0
Rian																-	30	30	30	05	50	0
Wilf																	-	50	45	25	50	0
Rosy																		-	45	25	50	0
Lady Di																			-	10	50	0
Paka																				-	0	0
Hans																					-	0
Nicholas																						-

Table E.3   Probable coefficient of relatedness between the members of the Kaspersdraai spotted hyaena clan.

Underlined individuals are females.

	Marlin	Agfa	Cyclops	Sally	Patrick	Prince Valliant	Blommetjie	Harrison	Vega
Marlin	-	50	0	50	50	0	25	25	50
Agfa		-	50	50	50	0	50	50	25
Cyclops			-	50	50	0	0	0	0
Sally				-	50	0	13	13	25
Patrick					-	0	13	13	25
Prince Valliant						-	50	50	50
Blommetjie							-	50	38
Harrison								-	38
Vega									-

References

Acocks, J. P. H. 1975. *Veld types of South Africa.* Botanical Survey of South Africa Memoir No.40, 2nd edn.

Andelt, W. F. 1987. Coyote predation. In *Wild furbearer management and conservation in North America*, M. Novak, J. A. Baker, M. E. Obbard & B. Malloch (eds), 128–40. Toronto: Ontario Ministry of Natural Resources.

Andersson, M. & J. Krebs 1978. On the evolution of hoarding behaviour. *Animal Behaviour* **26**, 707–11.

Anon. 1965. *Agricultural Economics Map of R.S.A.* Government Printer, Pretoria.

Anon. 1986. *Climate of South Africa: climate statistics up to 1984.* Weather Bureau, Pretoria.

Apps, P. J. 1982. Possible use of shamming by a brown hyaena in an aggressive encounter with a pride of lions. *South African Journal of Zoology* **17**, 91.

Attwell, C. A. M. 1980. Age determination of the blue wildebeest *Connocahetes taurinus* in Zululand. *South African Journal of Zoology* **15**, 121–30.

Barnard, B. J. H. 1979. The role played by wildlife in the epizootiology of rabies in South Africa and South West Africa. *Onderstepoort Journal of Veterinary Research* **46**, 155–63.

Bearder, S. K. 1975. Inter-relationships between hyaenas and their competitors in the Transvaal Lowveld. *Publikasie van die Universiteit van Pretoria. Nuwe Reeks* **97**, 39–47.

Bearder, S. K. 1977. Feeding habits of spotted hyaenas in woodland habitat. *East African Wildlife Journal* **14**, 233–35.

Bearder, S. K. & R. M. Randall 1978. The use of faecal marking sites by spotted hyaenas and civets. *Carnivore* **1**, 32–48.

Bekoff, M. & J. A. Byers 1985. The development of behaviour from evolutionary and ecological perspectives in mammals and birds. *Evolutionary Biology* **19**, 215–86.

Bekoff, M., T. J. Daniels & J. L. Gittleman 1984. Life history patterns and the comparative social ecology of carnivores. *Annual Review of Ecological Systems* **15**, 191–252.

Bekoff, M. & M. C. Wells 1980. The social ecology of coyotes. *Scientific American* **242**, 130–51.

Berg, W. E. & R. A. Chesness 1978. Ecology of coyotes in northern Minnesota. In *Coyotes: biology, behavior and management*, M. Bekoff (ed.), 229–47. New York: Academic Press.

Bertram, B. C. R. 1975. Social factors influencing reproduction in lions. *Journal of Zoology* **177**, 463–82.

Bertram, B. C. R. 1976. Kin selection in lions and in evolution. In *Growing Points in Ethology*, P. P. G. Bateson & R. A. Hinde (eds), 281–301. Cambridge: Cambridge University Press.

Bertram, B. C. R. 1979. Serengeti predators and their social systems. In *Serengeti. Dynamics of an ecosystem*, A. R. E. Sinclair & M. Norton-Griffiths (eds), 221–48. Chicago: Chicago University Press.

Bigalke, R. C. 1972. Observations on the behaviour and feeding habits of the springbok *Antidorcas marsupialis*. *Zoologica Africana* **7**, 333–59.

Blair Rains, A. & A. M. Yalala 1972. *The central and southern state lands, Botswana.* Land resource study No. 11. Land resources division, Surbiton, England.

Bothma, J. duP. 1966. Notes on the stomach contents of certain Carnivora (Mammalia) from the Kalahari Gemsbok National Park. *Koedoe* **9**, 37–39.

Bothma, J. duP. & E. A. N. le Riche 1984. Aspects of the ecology and behaviour of the leopard *Panthera pardus* in the Kalahari Desert. *Koedoe* **27** (supplement), 259–79.

Bothma, J. duP. & M. G. L. Mills 1977. Ungulate abundance in the Nossob river valley in the south-western Kalahari Desert. *Proceedings of XIII International Congress of Game Biologists, Atlanta*, 90–102.

Bowen, W. D. 1981. Variation in coyote social organization: the influence of prey size. *Canadian Journal of Zoology* **59**, 639–52.

Bowen, W. D. 1982. Home range and spatial organization of coyotes in Jasper National Park, Alberta. *Journal of Wildlife Management* **46**, 201–16.

Bradbury, J. W. & S. L. Vehrencamp 1976a. Social organisation and foraging in emballonurid bats. 1. Field studies. *Behavioural Ecology & Sociobiology* **1**, 337–81.

Bradbury, J. W. & S. L. Vehrencamp 1976b. Social organisation and foraging in emballonurid bats. II. A model for the determination of group size. *Behavioural Ecology & Sociobiology* **1**, 383–404.

Brooks, A. C. 1961. A study of the Thomson's gazelle (Gazella thomsonii Gunther) in Tanganyika. Tanganyika Colonial Research Publication No. 25.

Brooks, P. M., J. Hanks & J. V. Ludbrook 1977. Bone marrow as an index of condition in African ungulates. *South African Journal of Wildlife Research* **7**, 61–66.

Brown, J. L. and G. H. Orians 1970. Spacing patterns in mobile animals. *Annual Review of Ecology and Systems* **1**, 239–62.

Bygott, J. D., B. C. R. Bertram & J. P. Hanby 1979. Male lions in large coalitions gain reproductive advantages. *Nature* **282**, 839–41.

Camenzind, F. J. 1978. Behavioural ecology of the coyote in the National Elk Refuge, Jackson, Wyoming. In *Coyotes: biology, behavior and management*, M. Bekoff (ed.), 267–94. New York: Academic Press.

Caro, T. M. 1986a. The functions of stotting: a review of the hypotheses. *Animal Behaviour* **34**, 649–62.

Caro, T. M. 1986b. The functions of stotting in Thomson's gazelles: some test of the predictions. *Animal Behaviour* **34**, 663–84.

Carr, G. M. & D. W. Macdonald 1986. The sociality of solitary foragers: a model based on resource dispersion. *Animal Behaviour* **34**, 1540–49.

Chapman, R. C. 1978. Rabies: Decimation of a wolf pack in arctic Alaska. *Science* **201**, 365–67.

Cheatum, R. L. 1949. Bone marrow as an index of malnutrition in deer. *New York State Conservation* **3**, 19–22.

Chepko-Sade, B. D. & D. S. Sade 1979. Patterns of group splitting within matrilineal kinship groups. *Behavioural Ecology & Sociobiology* **5**, 67–86.

Clutton-Brock, T. H., F. E. Guinness & S. D. Albon 1982. *Red deer. Behaviour and ecology of two sexes*. Chicago: Chicago University Press.

Crandall, L. S. 1964. *The management of wild animals in captivity*. Chicago: Chicago University Press.

Dalke, P. D. 1942. The cottontail rabbits in Connecticut. *Bulletin of Connecticut State Geological and Natural History Survey No. 65*.

Davies, E. M. & P. D. Boersma 1984. Why lionesses copulate with more than one male. *The American Naturalist* **123**, 594–611.

Davies, N. B. 1978. Ecological questions about territorial behaviour. In *Behavioural ecology. An evolutionary approach*, J. R. Krebs & N. B. Davies (eds), 317–50. Oxford: Blackwell Scientific Publications.

De Groot, P. 1980. Information transfer in a socially roosting weaver bird (*Quelea quelea*: Ploceinae): an experimental study. *Animal Behaviour* **28**, 1249–54.

Deane, N. N. 1962. The spotted hyaena Crocutac crocuta. *Lammergeyer* **2**, 26–44.

Drickamer, L. C. 1974. A ten year summary of reproductive data for free-ranging *Macaca mulatta*. *Folia Primatologica* **21**, 61–80.

Dunbar, R. I. M. & E. P. Dunbar 1977. Dominance and reproductive success among female gelada baboons. *Nature* **266**, 351–56.

Eaton, R. L. 1981. The ethology, propagation and husbandry of the brown hyaena (*Hyaena brunnea*). *Zoologische Garten* **51**, 123–49.

Eisner, T. & J. A. Davis 1967. Mongoose throwing and smashing millipedes. *Science* **155**, 577–79.

Eloff, F. C. 1964. On the predatory habits of lions and hyaenas. *Koedoe* **7**, 105–13.

Eloff, F. C. 1966. Range extension of the blue wildebeest. *Koedoe* **9**, 34–36.

Eloff, F. C. 1973a. Ecology and behaviour of the Kalahari lion. In *The World's Cats*, R. L. Eaton (ed.), 90–126. Winston: World Wildlife Safari.

Eloff, F. C. 1973b. Lion predation in the Kalahari Gemsbok National Park. *Journal of South African Wildlife Management Association* **3**, 59–64.

Eloff, F. C. 1980. Cub mortality in the Kalahari lion *Panthera leo vernayi* (Roberts, 1948). *Koedoe* **23**, 163–70.

Eloff, F. C. 1984. Food ecology of the Kalahari lion *Panthera leo verhayi*. *Koedoe* **27** (supplement), 249–58.

Emlen, S. T. 1984. Cooperative breeding in birds and mammals. In *Behavioural ecology. An evolutionary approach*, 2nd edn, J. R. Krebs & N. B. Davies (eds), 305–39. Oxford: Blackwell Scientific Publications.

Ewer, R. F. 1954. Some adaptive features in the dentition of hyaenas. *Annals of the Magazine of Natural History* **7**, 188–94.

Ewer, R. F. 1967. The fossil hyaenids of Africa – a reappraisal. In *Background to evolution in Africa*, W. W. Bishop & J. D. Clark (eds), 109–23. Chicago: Chicago University Press.

Ewer, R. F. 1973. *The carnivores*. London: Weidenfeld & Nicolson.

Fitzgibbon, C. D. & J. H. Fanshawe 1988. Stotting in Thomson's gazelles: an honest signal of condition. *Behavioural Ecology and Sociobiology* **23**, 69–74.

Frame, G. W. & L. H. Frame 1981. *Swift and enduring*. New York: E. P. Dutton.

Frame, L. H. & G. W. Frame 1976. Female African wild dogs emigrate. *Nature* **263**, 227–29.

Frame, L. H., J. R. Malcolm, G. W. Frame & H. van Lawick 1979. Social organization of African wild dogs (*Lycaon pictus*) on the Serengeti Plains, Tanzania. *Zeitschrift für Tierpsychologie* **50**, 225–49.

Frank, L. G. 1986a. Social organization of the spotted hyaena (*Crocuta crocuta*). I. Demography. *Animal Behaviour* **35**, 1500–09.

Frank, L. G. 1986b. Social organization of the spotted hyaena (*Crocuta crocuta*). II. Dominance and reproduction. *Animal Behaviour* **35**, 1510–27.

Frank, L. G., J. M. Davidson & E. R. Smith 1985. Androgen levels in the spotted hyaena *Crocuta crocuta*: the influence of social factors. *Journal of Zoology* **206**, 525–31.

Fuller, T. K. & L. B. Keith 1981. Non-overlapping ranges of coyotes and wolves in northeastern Alberta. *Journal of Mammalogy* **62**, 403–5.

Giles, R. H. G. 1978. *Wildlife management*. San Franscisco: W. H. Freeman.

Gittleman, J. L. & P. H. Harvey 1982. Carnivore home-range size, metabolic needs and ecology. *Behavioural Ecology & Sociobiology* **10**, 57–63.

Golding, R. R. 1969. Birth and development of spotted hyaenas at the University of Ibadan zoo, Nigeria. *International Zoo Year Book* **9**, 93–95.

Gorman, M. L. 1989. Olfactory communication in carnivores. In *Carnivore behaviour, ecology and evolution*. J. Gittleman (ed.), 57–88. Ithaca N.Y.: Cornell University Press.

Gorman, M. L. & M. G. L. Mills 1984. Scent marking strategies in hyaenas (Mammalia). *Journal of Zoology* **202**, 535–47.

Gosling, M. 1982. A reassessment of the function of scent marking in territories. *Zeitschrift für Tierpsychologie* **60**, 89–118.

Goss, R. A. 1986. *The influence of food source on the behavioural ecology of brown hyaenas Hyaena brunnea in the Namib Desert*. MSc. thesis, University of Pretoria.

Gould, S. J. & E. S. Vrba 1982. Exaptation – a missing term in the science of form. *Paleobiology* **8**, 4–15.

Green, B., J. Anderson & T. Whateley 1984. Water and sodium turnover and estimated food consumption in free-living lions (*Panthera leo*) and spotted hyaenas (*Crocuta crocuta*). *Journal of Mammalogy* **65**, 593–99.

Grobler, J. H. 1980. Body growth and age determination of the sable *Hippotragus niger niger* (Harris, 1838). *Koedoe* **23**, 131–56.

Gustavson, C. R., J. Garcia, W. G. Hankins & K. W. Rusiniak 1974. Coyote predation control by aversive conditioning. *Science* **184**, 581–83.

Gustavson, C. R. & L. K. Nicolaus 1987. Taste aversion conditioning in wolves, coyotes, and other canids: Retrospect and prospect. In *Man and Wolf*, H. Frank (ed.), 169–200. Dordrecht: W. Junk.

Halliday, T. R. 1978. Sexual selection and mate choice. In *Behavioural ecology. An evolutionary approach*, J. R. Krebs & N. B. Davies (eds), 180–213. Oxford: Blackwell Scientific Publications.

Hamilton, W. J., R. L. Tilson & L. G. Frank 1986. Sexual monomorphism in spotted hyaenas, *Crocuta crocuta*. *Ethology* **71**, 63–73.

Hanby, J. P. & J. D. Bygott 1987. Emigration of subadult lions. *Animal Behaviour* **35**, 161–69.

Harrington, F. H., L. D. Mech & S. H. Fritts 1983. Pack size and wolf pup survival: their relationship under varying ecological conditions. *Behavioural Ecology and Sociobiology* **13**, 19–26.

Hendey, Q. B. 1974. New fossil canivores from the Swartkrans australopithecine site (Mammalia: Carnivora). *Annals of the Transvaal Museum* **29**, 27–48.

Henschel, J. R. 1986. *The socio-ecology of a spotted hyaena* Crocuta crocuta *clan in the Kruger National Park*. DSc. thesis, University of Pretoria.

Henschel, J. R. & J. D. Skinner 1987. Social relationships and dispersal patterns in a clan of spotted hyaenas *Crocuta crocuta* in the Kruger National Park. *South African Journal of Zoology* **22**, 18–24.

Henschel, J. R. & R. L. Tilson 1988. How much does a spotted hyaena eat? Perspective from the Namib Desert. *African Journal of Ecology* **26**, 247–55.

Henschel, J. R., R. Tilson and F. von Blottniz 1979. Implications of a spotted hyaena bone assemblage in the Namib Desert. *South African Archeological Bulletin* **34**, 127–31.

Herero, S. 1972. Bears: their biology and management. *IUCN Publications New Series* No. 23. Morges: Switzerland.

Hersteinsson, P. & D. W. Macdonald 1982. Some comparisons between red and arctic foxes, *Vulpes vulpes* and *Alopex lagopus* as revealed by radio tracking. *Symposium of Zoological Society London* **49**, 259–89.

Hill, A. 1980. Hyaena provisioning of juvenile offspring at the den. *Mammalia* **44**, 594–95.

Hitchins, P. M. & J. L. Anderson 1983. Reproduction, population characteristics and management of the black rhinoceros *Diceros bicornis minor* in the Hluhluwe/Corridor/Umfolozi Game Reserve Complex. *South African Journal of Wildlife Research* **13**, 78–85.

Houston, D. C. 1974. The role of griffon vultures as scavengers. *Journal of Zoology* **172**, 35–46.

Howell, F. C. & G. Petter 1980. The *Pachycrocuta* and *Hyaena* lineages (Plio-Pleistocene and extant species of the Hyaenidae). Their relationships with Miocene ictitheres: *Palhyaena* and *Hyaenictitherium*. *Geobios* **13**, 579–623.

Hrdy, S. B. 1979. Infanticide among animals: a review, classification, and examination of the implications for the reproductive strategies of females. *Ethology & Sociobiology* **1**, 13–40.

Huntley, B. J. 1974. Outlines of wildlife conservation in Angola. *Journal of South African Wildlife Management Association* **4**, 157–66.

Jeffery, R. C. V. & J. Hanks 1981. Age determination of eland *Taurotragus oryx* (Pallas, 1766) in the Natal highveld. *South African Journal of Zoology* **16**, 113–22.

Joubert, E. & P. K. N. Mostert 1975. Distribution patterns and status of some mammals in South West Africa. *Madoqua* 9, 5–44.

Kaplan, C. 1977. The world problem. In *Rabies: the facts*, C. Kaplan (ed.), 1–21. Oxford: Oxford University Press.

Kaufmann, J. H. 1983. On the definitions and functions of dominance and territoriality. *Biological Review* 58, 1–20.

Kerr, M. A. & H. H. Roth 1970. Studies on the agricultural utilization of semi-domesticated eland (*Taurotragus oryx*) in Rhodesia. 3. Horn development and tooth eruption as indicators of age. *Rhodesian Journal of Agricultural Research* 8, 149–55.

Keverne, E. B. 1985. Reproductive behaviour. In *Reproductive fitness*, C. R. Austin & R. V. Short (eds), 133–75. Cambridge: Cambridge University Press.

King, L. C. 1963. *South African scenery*, 3rd edn. Edinburgh: Oliver & Boyd.

Kleiman, D. 1967. Some aspects of social behaviour in the Canidae. *American Zoologist* 7, 365–72.

Kleiman, D. G. & J. F. Eisenberg 1973. Comparisons of canid and felid social systems from an evolutionary perspective. *Animal Behaviour* 21, 637–59.

Krebs, J. R. 1974. Colonial nesting and social feeding as strategies for exploiting food resources in the great blue heron (*Ardea herodias*). *Behaviour* 51, 99–134.

Krebs, J. R. 1978. Optimal foraging: Decision rules for predators. In *Behavioural Ecology. An evolutionary approach*, J. R. Krebs & N. B. Davies (eds), 23–63. Oxford: Blackwell.

Kruuk, H. 1964. Predators and anti-predator behaviour of the black-headed gull (*Larus ridibundus* L.). *Behaviour*, supplement 11, 1–130.

Kruuk, H. 1966. Clan-system and feeding habits of spotted hyaenas (*Crocuta crocuta* Erxleben). *Nature* 209, 1257–58.

Kruuk, H. 1970. Interactions between populations of spotted hyaenas (*Crocuta crocuta* Erxleben) and their prey species. In *Animal populations in relation to their food resources*, A. Watson (ed.), 359–74. Oxford: Blackwell.

Kruuk, H. 1972. *The spotted hyaena. A study of predation and social behaviour*. Chicago: The University of Chicago Press.

Kruuk, H. 1975a. Functional aspects of social hunting by carnivores. In *Function and evolution in behaviour*, G. Baerends, C. Beer & A. Manning (eds), 119–41. Oxford: Clarendon Press.

Kruuk, H. 1975b. *Hyaena*. Oxford: Oxford University Press.

Kruuk, H. 1976. Feeding and social behaviour of the striped hyaena (*Hyaena vulgaris* Desmarest). *East African Wildlife Journal* 14, 91–111.

Kruuk, H. 1980. *The effects of large carnivores on livestock and animal husbandry in Marsabit District, Kenya*. PAL Technical Report E-4, UNEP-MAB.

Kruuk, H., M. L. Gorman & A. Leitch 1984. Scent marking with the subcaudal gland by the European badger, *Meles meles*. *Animal Behaviour* 32, 899–907.

Kruuk, H. & D. W. Macdonald 1985. Group territories of carnivores: empires and enclaves. In *Behavioural ecology. Ecological consequences of adaptive behaviour*, R. Sibly & R. Smith (eds), 521–36. Oxford: Blackwell Scientific Publications.

Kruuk, H. & T. Parish 1982. Factors affecting population density, group size and territory size of the European badger, *Meles meles*. *Journal of Zoology* 196, 31–39.

Kruuk, H. & T. Parish 1987. Changes in the size of groups and ranges of the European badger (*Meles meles* L.) in an area in Scotland. *Journal of Animal Ecology* 56, 351–64.

Kruuk, H. & W. A. Sands 1972. The aardwolf (*Proteles cristatus* Sparrman 1783) as predator of termites. *East African Wildlife Journal* 10, 211–27.

Labuschagne, W. 1979. '*n Bio-ekologiese en gedragstudie van die jagluiperd* Acinonyx jubatus jubatus (*Schreber, 1776*). MSc. thesis, University of Pretoria.

Lamprecht, J. 1978. The relationship between food competition and foraging group size in some larger carnivores. *Zeitschrift für Tierpsychologie* 46, 337–43.

Lamprecht, J. 1981. The function of social hunting in larger terrestrial carnivores. *Mammal Review* 11, 169–79.

Lancaster, I. N. 1979. Evidence for a widespread late Pleistocene humid period in the Kalahari. *Nature* 279, 145–46.

Land, R. B. 1985. Genetics and reproduction. In *Reproduction fitness*, C. R. Austin & R. V. Short (eds), 24–61. Cambridge: Cambridge University Press.

Lang, Von E. M. 1958. Zur Haltung des Strandwolfes (*Hyaena brunnea*). *Der Zoologische Garten* 24, 81–90.

Leistner, O. A. 1959a. Notes on the vegetation of the Kalahari Gemsbok National Park with special reference to its influence on the distribution of antelopes. *Koedoe* 2, 128–51.

Leistner, O. A. 1959b. Preliminary list of plants found in the Kalahari Gemsbok National Park. *Koedoe* 2, 152–72.

Leistner, O. A. 1967. *The plant ecology of the southern Kalahari*. Botanical Survey of South Africa Memoir No. 38.

Leistner, O. A. & M. J. A. Werger 1973. Southern Kalahari phytosociology. *Vegetatio* 28, 353–97.

Lindeque, M. 1981. *Reproduction in spotted hyaena* Crocuta crocuta (*Erxleben*). MSc thesis, University of Pretoria.

Lindeque, M. & J. D. Skinner 1982a. Aseasonal breeding in the spotted hyaena (*Crocuta crocuta*, Erxleben) in southern Africa. *African Journal of Ecology* 20, 271–78.

Lindeque, M. & J. D. Skinner 1982b. Fetal androgens and sexual mimicry in spotted hyaenas (*Crocuta crocuta*). *Journal of Reproductive Fertility* 65, 405–10.

Macdonald, D. W. 1976. Food caching by red foxes and some other carnivores. *Zeitschrift für Tierpsychologie* 42, 170–85.

Macdonald, D. W. 1980a. Social factors affecting reproduction amongst red foxes (*Vulpes vulpes* L., 1758). In *The red fox: Symposium on behaviour and ecology*, E. Zimen (ed.), 131–83. The Hague: Junk.

Macdonald, D. W. 1980b. *Rabies and wildlife, a biologist's perspective*. Oxford: Oxford University Press.

Macdonald, D. W. 1981. Resource dispersion and the social organisation of the red fox (*Vulpes vulpes*). In *Worldwide Furbearer Conference Proceedings*, Vol. 2, J. Chapman and D. Pursley (eds), 918–49. Virginia: R. R. Donnelley.

Macdonald, D. W. 1983. The ecology of carnivore social behaviour. *Nature* 301, 379–84.

Macdonald, D. W. & P. J. Bacon 1982. Fox society, contact rate and rabies epizootiology. *Comparative Immunology, Microbiology and Infectious Diseases* 5, 247–56.

Macdonald, D. W., F. G. Ball & N. G. Hough 1980. The evaluation of home range size and configuration using radio tracking data. In *A handbook on biotelemetry and radio tracking*, C. J. Amlaner & D. W. Macdonald (eds), 387–404. Oxford: Pergamon Press.

Macdonald, D. W. & L. Boitani 1979. The management and conservation of carnivores: a plea for an ecological ethic. In *Animal Rights*, D. Patterson & R. Ryder (eds), 165–77. Cambridge: Centaur Press.

Macdonald, D. W. & P. D. Moehlman 1982. Cooperation, altruism and restraint in the reproduction of carnivores. In *Perspectives in ethology*, P. P. G. Bateson & P. Klopfer (eds), 433–66. New York: Plenum Press.

Majerus, M. E. N. 1986. The genetics and evolution of female choice. *Trends in Ecology and Evolution* 1, 1–7.

Malcolm, J. R. and K. Marten 1982. Natural selection and the communal rearing of pups in African wild dogs (*Lycaon pictus*). *Behavioural Ecology & Sociobiology* 10, 1–13.

Matthews, L. H. 1939. Reproduction in the spotted hyaena, *Crocuta crocuta* (Erxleben). *Philosophical Transactions* Series B 230, 1–78.

Maynard Smith, J. & G. A. Parker 1976. The logic of asymmetric contests. *Animal Behaviour* 24, 159–75.

Mech, L. D. 1970. *The wolf: the ecology and behaviour of an endangered species.* New York: The Natural History Press.

Mech, L. D. 1977. A recovery plan for the eastern timber wolf. *National Parks and Conservation Magazine* January, 17–21.

Meissner, H. H. 1982. Classification of farm and game animals to predict carrying capacity. *Farming in South Africa, Wool Production C.3.* Pretoria: Department Agriculture and Fisheries.

Meredith, C. D. 1982. Wildlife rabies: past and present in South Africa. *South African Journal of Science* 78, 411–15.

Mills, M. G. L. 1978a. Foraging behaviour of the brown hyaena (*Hyaena brunnea* Thunberg, 1820) in the southern Kalahari. *Zeitschrift für Tierpsychologie* 48, 113–41.

Mills, M. G. L. 1978b. The comparative socio-ecology of the hyaenidae. *Carnivore* 1, 1–7.

Mills, M. G. L. 1982a. Factors affecting group size and territory size of the brown hyaena, *Hyaena brunnea*, in the southern Kalahari. *Journal of Zoology* 198, 39–51.

Mills, M. G. L. 1982b. The mating system of the brown hyaena in the southern Kalahari. *Behavioural Ecology and Sociobiology* 10, 131–36.

Mills, M. G. L. 1982c. Notes on age determination, growth and measurements of brown hyaenas *Hyaena brunnea* from the Kalahari Gemsbok National Park. *Koedoe* 25, 55–62.

Mills, M. G. L. 1983a. Behavioural mechanisms in territory and group maintenance of the brown hyaena, *Hyaena brunnea*, in the southern Kalahari. *Animal Behaviour* 31, 503–10.

Mills, M. G. L. 1983b. Mating and denning behaviour of the brown hyaena and comparisons with other Hyaenidae. *Zeitschrift für Tierpsychologie* 63, 331–42.

Mills, M. G. L. 1984a. The comparative behavioural ecology of the brown hyaena *Hyaena brunnea* and the spotted hyaena *Crocuta crocuta* in the southern Kalahari. *Koedoe* 27 (supplement), 237–47.

Mills, M. G. L. 1984b. Prey selection and feeding habits of the large carnivores in the southern kalahari. *Koedoe* 27 (supplement), 281–94.

Mills, M. G. L. 1985a. Related spotted hyaenas forage together but do not cooperate in rearing young. *Nature* 316, 61–62.

Mills, M. G. L. 1985b. Hyaena survey of Kruger National Park: August–October 1984. *I.U.C.N. S.S.C. Hyaena Specialist Group Newsletter* No. 2, 15–25.

Mills, M. G. L. 1989. The comparative behavioral ecology of hyenas: the importance of diet and food dispersion. In *Carnivore behavior, ecology and evolution*, J. L. Gittleman (ed.), 125–42. Ithaca, N.Y.: Cornell University Press.

Mills, M. G. L. & M. L. Gorman 1987. The scent-marking behaviour of the spotted hyaena *Crocuta crocuta* in the southern Kalahari. *Journal of Zoology* 212, 483–97.

Mills, M. G. L., M. L. Gorman & M. E. J. Mills 1980. The scent marking behaviour of the brown hyaena *Hyaena brunnea*. *South African Journal of Zoology* 15, 240–48.

Mills, M. G. L. & M. E. J. Mills 1977. An analysis of bones collected at hyaena dens in the Kalahari Gemsbok National Parks (Mammalia: Carnivora). *Annals of the Transvaal Museum* 30, 145–55.

Mills, M. G. L. & M. E. J. Mills 1978. The diet of the brown hyaena *Hyaena brunnea* in the southern Kalahari. *Koedoe* 21, 125–49.

Mills, M. G. L. & M. E. J. Mills 1982. Factors affecting the movement patterns of brown hyaenas, *Hyaena brunnea*, in the southern Kalahari. *South African Journal of Wildlife Research* 12, 111–17.

Mills, M. G. L. & P. F. Retief 1984a. The response of ungulates to rainfall along the riverbeds of the southern Kalahari, 1972–1982. *Koedoe* 27 (supplement), 129–42.

Mills, M. G. L. & P. F. Retief 1984b. The effect of windmill closure on the movement patterns of ungulates along the Auob riverbed. *Koedoe* 27 (supplement), 107–10.

Mills, M. G. L., P. Wolff, E. A. N. le Riche & I. J. Meyer 1978. Some population characteristics of the lion (*Panthera leo*) in the Kalahari Gemsbok National Park. *Koedoe* 21, 163–71.

Moehlman, P. D. 1979. Jackal helpers and pup survival. *Nature* 277, 382–3.

Moehlman, P. D. 1981. Reply to R. D. Montgomerie. *Nature* **289**, 825.

Moehlman, P. D. 1983. Socioecology of silverbacked and golden jackals, *Canis mesomelas* and *C. aurens*. In *Recent Advances in the Study of Mammalian Behaviour*, J. F. Eisenberg & D. G. Kleiman (eds), 423–53. Lawrence, Kans.: American Society of Mammalogists Special Publication No. 7.

Mohr, C. O. 1947. Table of equivalent populations of North American small mammals. *American Midland Naturalist* **37**, 223–49.

Mundy, P. J. & J. A. Ledger 1976. Griffon vultures, carnivores and bones. *South African Journal of Science* **72**, 106–10.

Nash, L. T. 1976. Troop fission in free-ranging baboons in the Gombe Stream National Park, Tanzania. *Journal of Physical Anthropology* **44**, 63–68.

Owens, D. D. & M. J. Owens 1979a. Communal denning and clan associations in brown hyaenas (*Hyaena brunnea*, Thunberg) of the Central Kalahari Desert. *African Journal of Ecology* **17**, 35–44.

Owens, D. D. & M. J. Owens 1979b. Notes on social organization and behaviour in brown hyaenas (*Hyaena brunnea*). *Journal of Mammalogy* **60**, 405–8.

Owens, D. D. & M. J. Owens 1984. Helping behaviour in brown hyaenas. *Nature* **308**, 843–45.

Owens, M. J. & D. D. Owens 1978. Feeding ecology and its influence on social organisation in brown hyaenas (*Hyaena brunnea*, Thunberg) of the Central Kalahari Desert. *East African Wildlife Journal* **16**, 113–36.

Packer, C. 1983. Sexual dimorphism. The horns of African antelopes. *Science* **221**, 1191–93.

Packer, C. 1986. The ecology of sociality in Felids. In *Ecological aspects of social evolution*, D. I. Rubenstein & R. W. Wrangham (eds), 429–51. Princeton, NJ: Princeton University Press.

Packer, C., L. Herbst, A. E. Pusey, J. D. Bygott, J. P. Hanby, S. J. Cairns & M. Borgerhoff-Mulder 1988. Reproductive success of lions. In *Reproductive success*, T. H. Clutton-Brock (ed.), 363–83. Chicago: University of Chicago Press.

Packer, C. & A. E. Pusey 1982. Cooperation and competition within coalitions of male lions: kin selection or game theory? *Nature* **296**, 740–42.

Packer, C. & A. E. Pusey 1983a. Male takeovers and female reproductive parameters: a simulation of oestrous synchrony in lions (*Panthera leo*). *Animal Behaviour* **31**, 334–40.

Packer, C. & A. E. Pusey 1983b. Adaptations of female lions to infanticide by incoming males. *The American Naturalist* **121**, 716–28.

Partridge, L. & T. R. Halliday 1984. Mating patterns and mate choice. In *Behavioural ecology. An evolutionary approach*, 2nd edn, J. R. Krebs & N. B. Davies (eds), 222–50. Oxford: Blackwell Scientific Publications.

Patterson, I. J. 1965. Timing and spacing of broods in the black-headed gull. *Ibis* **107**, 433–59.

Peters, R. P. & L. D. Mech 1975. Scent marking in wolves. *American Scientist* **63**, 628–37.

Peterson, R. O. 1977. Wolf ecology and prey relationships on Isle Royale. *National Parks Service Scientific Monograph Series* No. 11.

Peterson, R. O., J. D. Woolington & T. N. Bailey 1984. Wolves of the Kenai Peninsula, Alaska. *Wildlife Monographs*, 88.

Pettifer, H. L. 1981a. The experimental release of captive-bred cheetah (*Acinonyx jubatus*) into the natural environment. In *Worldwide Furbearer Conference Proceedings*, Vol. 1, J. Chapman & D. Pursley (eds), 1001–24. Virginia: R. R. Donnelley.

Pettifer, H. L. 1981b. Aspects of the ecology of cheetahs (*Acinonyx jubatus*) on the Suikerbosrand Nature Reserve. In *Worldwide Furbearer Conference Proceedings*, Vol. 2, J. A. Chapman & D. Pursley (eds), 1121–42. Virginia: R. R. Donnelley.

Pienaar, U. deV. 1969. Predator–prey relationships amongst the larger mammals of the Kruger National Park. *Koedoe* 12, 108–76.

Pond, C. M. 1977. The significance of lactation in the evolution of mammals. *Evolution* 31, 177–99.

Pournelle, G. H. 1965. Observations on birth and early development of the spotted hyaena. *Journal of Mammalogy* 46, 503.

Pusey, A. E. & C. Packer 1987. The evolution of sex-based dispersal in lions. *Behaviour* 101, 275–310.

Racey, P. A. & J. D. Skinner 1979. Endocrine aspects of sexual mimicry in spotted hyaenas *Crocuta crocuta*. *Journal of Zoology* 187, 315–26.

Ralls, K. 1976. Mammals in which females are larger than males. *Quarterly Review of Biology* 51, 245–76.

Rasa, O. A. E. 1987. The dwarf mongoose: A study of behaviour and social structure in relation to ecology in a small, social carnivore. In *Advances in the study of behaviour* 17, J. S. Rosenblatt & P. Slater (eds), 121–60. New York: Academic Press.

Rautenbach, I. L. 1971. Ageing criteria in the springbok *Antidorcas marsupialis* (Zimmermann, 1780) (Artiodactyla: Bovidae). *Annals of the Transvaal Museum* 27, 83–133.

Reich, A. 1981. *The behaviour and ecology of the African wild dog (*Lycaon pictus*) in the Kruger National Park*. PhD thesis, Yale University.

Richards, S. M. 1974. The concept of dominance and methods of assessment. *Animal Behaviour* 22, 914–30.

Richardson, P. R. K. 1985. *The social behaviour and ecology of the aardwolf,* Proteles cristatus *(Sparrman, 1783) in relation to its food resources*. D.Phil. thesis, Oxford University.

Richardson, P. R. K. 1987. Aardwolf mating system: Overt cuckoldry in an apparently monogamous mammal. *South African Journal of Science* 83, 405–10.

Richardson, P. R. K., P. Mundy & I. Plug 1986. Bone crushing carnivores and their significance in the growth of griffon vulture chicks. *Journal of Zoology* 210, 23–43.

Rieger, I. 1979. Breeding the striped hyaena *Hyaena hyaena* in captivity. *International Zoo Year Book* 19, 193–98.

Rieger, I. 1981. *Hyaena hyaena*. *Mammalian Species* No. 150, 1–5.

Rood, J. P. 1983. The social system of the dwarf mongoose. In *Advances in the study of mammalian behaviour*, J. F. Eisenberg & D. G. Kleiman (eds), 454–88. Lawrence, Kans.: American Society of Mammalogists Special Publication No. 7.

Rowe-Rowe, D. T. 1973. Horn development of a grey rhebuck. *Lammergeyer* 19, 36–38.

Rowe-Rowe, D. T. 1978a. Comparative prey capture and food studies of South African mustelines. *Mammalia* 42, 175–96.

Rowe-Rowe, D. T. 1978b. Reproduction and post-natal development of South African mustelines (Carnivora: Mustelidae). *Zoologica Africana* 13, 103–14.

Rowell, T. E. 1974. The concept of social dominance. *Behavioural Ecology* 11, 131–54.

Rudnai, J. 1973. Reproductive biology of lions (*Panthera leo massaica* Neumann) in Nairobi National Park. *East African Wildlife Journal* 11, 241–53.

Sampson, R. J. 1978. *Surface II graphics system*. Lawrence: Kansas Geological Survey.

Savage, R. J. G. 1978. Carnivora. In *Evolution of African Mammals*, V. J. Maglio & H. B. S. Cooke (eds), 249–67. Cambridge, Mass: Harvard University Press.

Schaller, G. B. 1967. *The deer and the tiger*. Chicago: Chicago University Press.

Schaller, G. B. 1972. *The Serengeti lion. A study of predator–prey relations*. Chicago: Chicago University Press.

Schneider, K. M. 1926. Über Hyaenenzucht. *Die Pelztierzucht* 2, 1–14.

Schultz, W. C. 1966. Breeding and hand-rearing of brown hyaenas at Okahandja Zoopark. *International Zoo Year Book* 6, 173–76.

Seal, U. S. 1969. Carnivore systematics: a study of haemoglobins. *Comparative Biochemistry and Physiology* **31**, 799–811.

Seidensticker, J. C. 1976. On the ecological separation between tigers and leopards. *Biotropica* **8**, 225–34.

Shoemaker, A. H. 1978. Studbook for the brown hyaena (*Hyaena brunnea*) in captivity. *International Zoo Year Book* **18**, 224–27.

Shoemaker, A. H. 1983. 1982 Studbook report on the brown hyaena, *Hyaena brunnea*: Decline of a pedigree species. *Zoo Biology* **2**, 133–36.

Siegal, S. 1956. *Non-parametric statistics for the behavioural sciences*. New York: McGraw-Hill.

Siegfried, W. R. 1984. An analysis of faecal pellets of the brown hyaena on the Namib coast. *South African Journal of Zoology* **19**, 61.

Sikes, S. K. 1964. The ratel or honey badger. *African Wild Life* **18**, 29–37.

Silk, J. B., C. B. Clark-Wheatley, P. Rodman & A. Samuels 1981. Differential reproductive success and facultative adjustment of sex ratios among captive female bonnet macaques (*Macaca radiata*). *Animal Behaviour* **29**, 1106–20.

Simpson, C. D. 1972. Some characteristics of Tragelaphine horn growth and their relationship to age in greater kudu and bushbuck. *Journal of South African Wildlife Management Association* **2**, 1–8.

Skinner, J. D. 1976. Ecology of the brown hyaena *Hyaena brunnea* in the Transvaal with a distribution map for southern Africa. *South African Journal of Science* **72**, 262–69.

Skinner, J. D. & G. Ilani 1979. The striped hyaena *Hyaena hyaena* of the Judean and Negev Deserts and a comparison with the brown hyaena *H. brunnea*. *Israel Journal of Zoology* **28**, 229–32.

Skinner, J. D. & R. J. van Aarde 1981. The distribution and ecology of the brown hyaena *Hyaena brunnea* and spotted hyaena *Crocuta crocuta* in the central Namib Desert. *Madoqua* **12**, 231–39.

Skinner, J. D., R. J. van Aarde & A. S. van Jaarsveld 1984. Adaptations in three species of large mammals (*Antidorcas marsupialis*, *Hystrix africaeaustralis*, *Hyaena brunnea*) to arid environments. *South African Journal of Zoology* **19**, 82–86.

Smithers, R. H. N. 1971. *The mammals of Botswana*. Salisbury: Trustees, National Museum, Rhodesia.

Smithers, R. H. N. 1983. *The mammals of the southern African subregion*. Pretoria: University of Pretoria.

Smithers, R. H. N. & J. L. P. Lobao Tello 1976. *Check list and atlas of the mammals of Mocambique*. Salisbury: Trustees, National Museums and Monuments, Rhodesia.

Smithers, R. H. N. & V. J. Wilson 1979. *Checklist and atlas of the mammals of Zimbabwe-Rhodesia*. Salisbury: Trustees, National Museums and Monuments, Zimbabwe-Rhodesia.

Smuts, G. L. 1979. Diets of lions and spotted hyaenas assessed from stomach contents. *South African Journal of Wildlife Research* **9**, 19–25.

Smuts, G. L., J. F. Anderson & J. C. Austin 1978. Age determination of the African lion. *Journal of Zoology* **185**, 115–46.

Smuts, G. L., G. A. Robinson & I. J. Whyte 1980. Comparative growth of wild male and female lions (*Panthera leo*). *Journal of Zoology* **190**, 365–73.

Smuts, G. L., I. J. Whyte & T. W. Dearlove 1977a. A mass capture technique for lions. *East African Wildlife Journal* **15**, 81–87.

Smuts, G. L., I. J. Whyte & T. W. Dearlove 1977b. Advances in the mass capture of lions (*Panthera leo*). *Proceedings XIII Intenational Congress of Game Biologists, Atlanta*, 420–31.

Sokal, R. R. & F. J. Rohlf 1969. *Biometry*. San Franscisco: W. H. Freeman.

Spinage, C. A. 1967. Ageing the Uganda defassa waterbuck *Kobus defassa ugandae* Neumann. *East African Wildlife Journal* **5**, 1–17.

Spinage, C. A. 1971. Geratodontology and horn growth of the impala (*Aepyceros melampus*). *Journal of Zoology* **164**, 209–25.

Sterner, R. T. & S. A. Shumaker 1978. Coyote damage-control research: A review and analysis. In *Coyotes: Biology, behaviour and management*, M. Bekoff (ed.), 297–325. New York: Academic Press.

Story, R. 1958. *Some plants used by bushmen in obtaining food and water.* Botanical Survey of South Africa, Memoir No. 30.

Stuart, C. T. 1976. Plant food in the diet of the spotted hyaena. *South African Journal of Science* **72**, 148.

Stuart, C. T., I. A. W. Macdonald & M. G. L. Mills 1985. History, current status and conservation of large mammalian predators in Cape Province, Republic of South Africa. *Biological Conservation* **31**, 7–19.

Stuart, C. T. & P. D. Shaughnessy 1984. Content of *Hyaena brunnea* and *Canis mesomelas* scats from southern coastal Namibia. *Mammalia* **48**, 611–12.

Talbot, L. M. & M. H. Talbot 1963. The wildebeest in Western Masailand, East Africa. *Wildlife Monographs*, 12.

Thenius, E. 1966. Zur Stammesgeschichte der Hyaenen (Carnivora, Mammalia). *Zeitschrift für Saugetierkunde* **31**, 293–300.

Tilson, R. L. & W. J. Hamilton 1984. Social dominance and feeding patterns of spotted hyaenas. *Animal Behaviour* **32**, 715–24.

Tilson, R. L. & J. R. Henschel 1986. Spatial arrangement of spotted hyaena groups in a desert environment, Namibia. *African Journal of Ecology* **24**, 173–80.

Tilson, R., F. von Blottnitz & J. Henschel 1980. Prey selection by spotted hyaenas (*Crocuta crocuta*) in the Namib Desert. *Madoqua* **12**, 41–9.

Tinbergen, N. 1965. Von den vorratskammern des Rotfuchses (*Vulpes vulpes* L.). *Zeitschrift für Tierpsychologie* **22**, 119–49.

Van der Walt, P. T., P. F. Retief, E. A. N. le Riche, M. G. L. Mills & G. de Graaff 1984. Features of habitat selection by larger herbivorous mammals and the ostrich in the southern Kalahari conservation areas. *Koedoe* **27** (supplement), 119–28.

Van Gelder, R. G. 1953. The egg-opening technique of a spotted skunk. *Journal of Mammalogy* **34**, 255–56.

Van Jaarsveld, A. S. & J. D. Skinner 1987. Spotted hyaena monomorphism: an adaptive 'phallusy'. *South African Journal of Science* **83**, 612–15.

Van Jaarsveld, A. S., J. D. Skinner & M. Lindeque 1988. Growth, development and parental investment in the spotted hyaena, *Crocuta crocuta*. *Journal of Zoology* **215**, 45–53.

Van Orsdol, K. G. 1981. Lion predation in Rwenzori National Park, Uganda. PhD thesis, Cambridge University.

Van Rooyen, N., D. J. van Rensberg, G. K. Theron & J. duP. Bothma 1984. A preliminary report on the dynamics of the vegetation of the Kalahari Gemsbok National Park. *Koedoe* **27** (supplement), 83–102.

Viljoen, P. J. 1980. *Veldtipes, verspreiding van die groter soogdiere, en enkele aspekte van die ekologie van Kaokoland.* MSc thesis, University of Pretoria.

Voigt, D. R. & R. R. Tinline 1980. Strategies for analysing radio tracking movement data. In *A handbook on biotelemetry and radio tracking*, C. J. Amlaner & D. W. Macdonald (eds), 387–404. Oxford: Pergamon Press.

Von Richter, W. 1972. Remarks on present distribution and abundance of some South African carnivores. *Journal of South African Wildlife Management Association* **2**, 9–16.

Von Richter, W. & T. Butynski 1973. Hunting in Botswana. *Botswana Notes and Records* **5**, 191–208.

Von Schantz, T. 1984. Spacing strategies, kin selection and population regulation in altricial vertebrates. *Oikos* **42**, 48–58.

Von Ullrich, F. & J. Schmitt 1969. Die Chromsomen des Erdwolfs, *Proteles cristatus* (Sparrman, 1893). *Zeitschrift für Saugetierkunde* **34**, 61–62.

Wade, D. A. 1978. Coyote damage: A survey of its nature and scope, control measures and their application. In *Coyotes: Biology, behaviour and management*, M. Bekoff (ed.), 347–68. New York: Academic Press.

Walther, F. R. 1969. Flight behaviour and avoidance of predators in Thomson's gazelle (*Gazella thomsonii* Gunther 1884). *Behaviour* **34**, 184–221.

Walther, F. R. 1984. *Communication and expression in hoofed mammals*. Bloomington: Indiana University Press.

Ward, P. & A. Zahavi 1973. The importance of certain assemblages of birds as 'information centres' for food finding. *Ibis* **115**, 517–34.

Watson, R. M. 1965. Observations on the behaviour of young spotted hyaena (*Crocuta crocuta*) in the burrow. *East African Wildlife Journal* **3**, 122–23.

Whateley, A. 1980. Comparative body measurements of male and female spotted hyaenas from Natal. *Lammergeyer* **28**, 40–43.

Whateley, A. 1981. Density and home range of spotted hyaenas in Umfolozi Game Reserve, Natal. *Lammergeyer* **31**, 15–20.

Whateley, A. & P. M. Brooks 1978. Numbers and movements of spotted hyaenas in Hluhluwe Game Reserve. *Lammergeyer* **26**, 44–52.

Wilson, E. O. 1975. *Sociobiology the new synthesis*. Cambridge & London: The Belknap Press of Harvard University Press.

Wolton, R. J. 1985. The ranging and nesting behaviour of wood mice, *Apodemus sylvaticus* (Rodentia: Muridae), as revealed by radio-tracking. *Journal of Zoology* **206**, 203–24.

Wrangham, R. W. 1980. An ecological model of female-bonded primate groups. *Behaviour* **71**, 262–99.

Wrangham, R. W. & D. I Rubenstein 1986. Social evolution in birds and mammals. In *Ecological aspects of social evolution*, D. I. Rubenstein & R. W. Wrangham (eds), 452–70. Princeton, NJ: Princeton University Press.

Wurster, D. H. 1969. Cytogenetic and phylogenetic studies in the order Carnivora. *Chromosoma* **24**, 336–82.

Wurster, D. H. & K. Benirschke 1968. Comparative cytogenetic studies in the order Carnivora. *Chromosoma* **24**, 336–82.

Wyman, J. 1967. The jackals of the Serengeti. *Animals* **10**, 79–83.

Yost, R. A. 1980. The nocturnal behaviour of captive brown hyaenas (*Hyaena brunnea*). *Mammalia* **44**, 27–34.

Index